THE SCARITH OF SCORNELLO

The Scarith of Scornello

A TALE OF RENAISSANCE FORGERY

Ingrid D. Rowland

The University of Chicago Press   Chicago *&* London

INGRID D. ROWLAND is the Andrew W. Mellon Professor in the Humanities at the American Academy in Rome. She is the author of *The Culture of the High Renaissance: Ancients and Moderns in Sixteenth-Century Rome* and *The Ecstatic Journey: Athanasius Kircher in Baroque Rome.*

The University of Chicago Press, Chicago 60637
The University of Chicago Press, Ltd., London
Printed in the United States of America

13 12 11 10 09 08 07 06 05   2 3 4 5

ISBN: 0-226-73036-0

Library of Congress Cataloging-in-Publication Data
Rowland, Ingrid D. (Ingrid Drake)
    The scarith of Scornello : a tale of Renaissance forgery / Ingrid D. Rowland.
        p. cm.
    Includes bibliographical references and index.
    ISBN 0-226-73036-0 (cloth : alk. paper)
    1. Inghirami, Curzio, 1614–1655. 2. Inghirami, Curzio, 1614–1655. Ethruscarum
antiquitatum fragmenta. 3. Literary forgeries and mystifications—History—
17th century. 4. Volterra (Italy)—History—17th century. 5. Tuscany (Italy)—
History—1434–1737. I. Title.
    DG975.V9R68 2004
    937'.5—dc22   2004012752

♾ The paper used in this publication meets the minimum requirements of the
American National Standard for Information Sciences—Permanence of Paper
for Printed Library Materials, ANSI Z39.48-1992.

FOR

Thomas Cerbu & Walter Kaiser

Contents

─────────────────────────┥ ✤ ┝─────────────────────────

Acknowledgments

This book would not have been possible without Thomas Cerbu, both for his incomparably erudite advice and his long friendship. Thanks also to Francesco Solinas, Luc Deitz, Ingo Herklotz, Paul Gehl, Andrew Butterfield, Andrea Galdy, Anthony Grafton, and Walter Stephens. It was Angelo Marrucci, the late librarian of the Biblioteca Comunale Guarnacci of Volterra, who first broke the news that Curzio Inghirami was not only a forger of Etruscan antiquities, but of much else. I owe the Etruscan side of the story to the late Kyle M. Phillips Jr., Ingrid E. M. Edlund, Françoise-Hélène Massa-Pairault, Mario Torelli, and Stephan Steingräber. Warm thanks to the staffs of the following libraries: Biblioteca Comunale Guarnacci, Volterra; Biblioteca Marucelliana, Florence; Biblioteca Laurenziana, Florence; Biblioteca Nazionale Centrale, Florence; Biblioteca Riccardiana, Florence; Biblioteca Apostolica Vaticana; Department of Special Collections, Getty Research Institute Library; Department of Special Collections at the Joseph M. Regenstein Library, University of Chicago. Thanks also to Franklin Spellman, antiquarian bookseller of Culver City, California, who put into my hands the most beautiful copy of Curzio Inghirami's *Ethruscarum Antiquitatum Fragmenta* that I have seen to date.

The whole problem of Curzio emerged from a larger project during a fellowship year at the Harvard Center for Renaissance Studies at Villa I Tatti in 1993–94. A fellowship at the Getty Research Institute in 2000–2001 helped give the story what seemed to be its final shape, but the story came to its rightful end only after I had made contact with the living Inghirami family; my thanks to Iacopo Inghirami, Chiara Bertini Inghirami, Carolina Inghirami, and Francesca Inghirami for their hospitality and their help in unraveling the mysteries of Scornello. Susan Bielstein at the University of Chicago Press saw potential in a strange story; Walter Stephens and Brian Curran acted as lively referees (culminating in Brian's marginal exultation "Allacci kicks it!"). Erin DeWitt edited with lynx eyes.

Walter Kaiser directed the Villa I Tatti at the time of my fellowship there, and without his incisive sense of humor, Curzio Inghirami's story would never have taken shape as a story with a happy ending. I dedicate this book to him and to his former student Thomas Cerbu.

THE SCARITH OF SCORNELLO

⟨ I ⟩

Discovery

1634

Cave, cave, cave
Beware, beware, beware.

On a beautiful afternoon in November 1634, lunch had ended
for the Inghirami family in their Tuscan villa. Curzio Inghirami,
aged nineteen (fig. 1), decided to go fishing with his thirteen-
year-old sister, Lucrezia, in the river behind their house.[1] They
were used to doing things together; their villa, Scornello (fig. 2),
stood on an isolated hill in the rugged countryside south of
Volterra, the highest and most remote of the great ancient Etruscan cities, and their only close neighbors were a few families
of tenant farmers.[2] Half an hour's walk brought them down to
the flat bed of the river Cecina, where the fishing was good that
afternoon. Curzio's manservant would later tell the police that
as they made their way back up the tree-lined road toward
home, brother and sister suddenly bent laughing over a "a mixture of certain hairs" (fig. 3). Curzio himself would tell the story

3

somewhat differently; he confessed that he had been standing on the riverbank, throwing rocks:

1 : Curzio Inghirami; from Lelio Lagorio, ed., *Dizionario di Volterra.*

It was when we were going fishing after lunch on Saint Catherine's Day, while, some three hundred cubits distant from our House, I looked over the river Cecina, waiting for the servants, and to amuse myself I rolled stones down the bank. It so happened that once another large stone had been moved aside, a small blackish clod was uncovered. . . . This I threw around several times, until it chanced to break apart. Then I could see that there were hairs underneath the layers of which the clod had been made. Astonished, I dissolved it with great effort. Many would come to think that the outer layer had been compacted of bitumen, pitch, resin, wax, frankincense, storax, and mastic, and other substances of this sort. The second layer, which was sturdier, was enveloped in fabric that had been mixed in with the hairs and reduced

2 : Scornello, Villa Inghirami and chapel. Author's photo.

Igitur cum ibidem ruri degeremus anno M. DC. XXXIV. & cum a prandio piscatum iremus die B. Catharinæ Virg. et Mart. dicato, dum longè à nostra Domo cubitos circiter tercentos Cecinam flumen inspicio, famulos expecto, & in decliue animi gratia saxa deuoluo, accidit, vt altro magno saxo emoto exiguus, & subniger globulus detegeretur huius formæ.

A
I.
Scarith
Die xxv.
Nouem-
bris.
MDCXXXV

B

Hic sapius à me coniectus tandem casu effringitur. Tum pilos, sub quibusdam corticibus, quibus ipse globulus compactus erat perspexi: his rebus in admirationem versus, ipsum magno cum labore dissolui. Primum corticem ex bitumine, pice, resina, cera, thure, storace, & mastice, alijsq́; huiusmodi compactum esse multi existimarunt. Secundus, qui validior erat, ijsdem pilis immixtus tela vinciebatur, quæ in puluerem comminuta est: sub illa quædam charta lintea hisce characteribus notata apparuit.

sorneth

P·ꝯ·A·

C

Ipsa verò aliam, quæ ob ictus infrusta abijt, continebat cum hoc vaticinio adamussim expresso.

Anno à Rege Iudæorum nunciato M. DC. XXIIII. Crucifixo M.D.XCI.
Veniet Canis fideliter seruiet seruitute libera jx. annos, & amplius.
Lupa mater Agni. Agnus amabit Canem. Veniet Porcus de grege Porcorum, & deuorabit labores Canis.

Caue,

3 : The first scarith, woodcut and movable type. Curzio Inghirami, *Ethruscarum Antiquitatum Fragmenta* (Florence: Massi, 1636), †ii 2v. Photo: Antonio Ortolan.

to powder, and underneath this there was linen rag paper marked with the following characters:

$$P \cdot 2I \cdot A \cdot$$

This paper contained another, which broke into fragments from the blows, with the following prophecy, precisely expressed:

In the year of the prophesied King of the Jews 1624, one thousand five hundred ninety-first from his Crucifixion:

A Dog shall come who shall serve out his term of indenture faithfully and freely for nine years, and more. The Wolf is the mother of the Lamb. The Lamb shall love the Dog. A Pig shall come forth from the horde of Pigs and shall devour the work of the Dog.

Cave, cave, cave [beware, beware, beware]

Prospero of Fiesole, resident of this colony, Guardian of the Citadel, Prophesied the year after Catilina's death. You have discovered the treasure. Mark the spot, and go away.[3]

Archaeological discoveries were nothing new around Volterra; the city was one of the oldest in Italy. Its imposing walls had been built in Etruscan times, nearly two thousand years before Curzio took his fishing trip. Patched and restored, they had served their purpose until 1472, when the city finally fell to Florence in a horrific siege (fig. 4).[4] Volterra's streets and buildings largely rested on Etruscan or Roman foundations, and where the city ended, limestone grave markers, marble statues, and tiny alabaster coffins came out of the earth with images of Volterrans long dead. Some of the names on these ancient graves were still familiar.[5] Plowmen turned up tiny Etrus-

4 : Etruscan walls of Volterra. Fototeca, American Academy in Rome.

can bronzes in their fields or stored their animals and equipment in Etruscan chamber tombs. Nothing, however, had ever looked quite like the capsule that Curzio Inghirami had just pulled from the riverbank.

Despite their portentous tone, almost comical in its pomposity, the prophecies recorded by Prospero the guardian also fit neatly into Curzio's experience of the world. Seventeenth-century Italians heard oracular predictions all the time from traveling preachers and passed them to one another by printed broadsheet, book, or word of mouth, much as their twentieth-century descendants would scan their newspapers for the horoscope and the weather report. The biblical book of Revelation provided a favorite source for these oracles, as did the prognostications of Joachim of Fiore, a Calabrian abbot who had first spouted similar meldings of apocalypse, political satire, and animal fable in the twelfth century. Prospero the guardian's fulminations about the lamb, the wolf, the dog, and the pig therefore belonged to a familiar form of speech, and not only for prophecies, but also for satires about political figures of the day, especially about Italy's uniquely peculiar politician, the pope.[6] The satires, especially, affected the same bombastic tones as Joachim's—or, now, Prospero's—prophecies, to comical effect.

However familiar the tone of Prospero's oracles, however, the guardian's final injunction to "mark the spot, and go away" sounded serious, and Curzio obeyed. In his own words: "I read it over, and I marveled; I marked the spot."[7] But the augur's scroll said nothing about souvenirs, so Curzio and Lucrezia took the whole strange bundle home with them, while their manservant went ahead with the fish; it had been, all in all, a productive afternoon. When they brought their package into the house, their grandmother spluttered, "It's a curse!" and ordered them to throw it away. Luckily their father, Inghiramo, took a more sober interest in the sturdy capsule with its paper scroll, so much so that he and Curzio returned to the riverbank with shovels the very next day—"You have discovered the treasure" seemed to impress Inghiramo more than the warning to "mark the spot, and go away." But their search for Prospero's treasure proved frustrating; the amateur archaeologists uncovered only a series of clay vessels that had been broken in antiquity, their contents long since carried off. That single day's taste of excavation was enough for Inghiramo, who took off for Florence shortly afterward, bound on his own business, which

included not only their farms and pastures, but also the salt springs that welled up on their property and at the base of the hill, one of the many oddities of Volterra's strange, mineral-rich terrain.[8]

There matters stood until 13 December, when the Inghirami's tutor, Father Domenico Vadorini, came by Scornello to celebrate the Feast of Saint Lucy with the family, say Mass in the villa's little chapel, and hear the children's lessons. When Curzio brought out the paper scroll with its Etruscan prophecy, Vadorini urged another try at excavation with a larger team of diggers. Together they summoned the only candidates available, a contingent of Scornello's tenant farmers, and marched back to the site to make their assault. Seasoned spade-workers, the *contadini* soon cleared a sturdy wall with a stone urn embedded in one of its crannies. The urn contained another bundle like the one Curzio had found by the river, again inscribed on the outside, and again with what appeared to be Etruscan letters (fig. 5).

Confirmation of the bundle's Etruscan origin was easy to secure. Volterra boasted a famous pair of Etruscan inscriptions, both unearthed in 1494 by the famous local scholar Raffaele Maffei, whose descendant, Raffaello Maffei, was Curzio Inghirami's best friend. A visit next day to the Palazzo Maffei gave Curzio and Father Vadorini the information they needed: the letters on this new capsule were almost identical to those on the ancient tomb markers.[9] Together with Raffaello Maffei and some of Curzio's other friends in town (the "more studious" among them, he recalled), they broke the capsule open with great ceremony and found to their delight that it, too, was stuffed with rolls of paper documents. This time, however, some texts were written in Etruscan (fig. 6) and some in Latin.

The Etruscan texts could not be read, but their content could be guessed from the Latin records: once again Curzio

II.
Scarith
Die xiij.
Decem-
bris.
M.DC.XXX
IV.

Hoc Vulterram defero, & cum marmoreis monumentis in Theatro Vulterrano à Raphaele Maffæo anno 1494 repertis confero, ciusdem formæ esse characteres comperio. Aperui autem coram plurimis Patriæ meæ eruditis adolescentibus, qui mecum, quas appono, delineatas Schedulas, Ethruscas, & Latinas venerati sunt.

5 : Scarith #2, woodcut. Curzio Inghirami, *Ethruscarum Antiquitatum Fragmenta* (Florence: Massi, 1636), †ii 3v. Photo: Antonio Ortolan.

6 : "Etruscan" texts from Scarith #2, copper-plate engraving above, woodcut below. Curzio Inghirami, *Ethruscarum Antiquitatum Fragmenta* (Florence: Massi, 1636), ††r. Photo: Antonio Ortolan.

learned that the cache of documents was connected with the figure of Prospero of Fiesole, who could now be identified as a novice in training to become an augur, an Etruscan priest:

> I, Prospero, was instructed in the art of divination by my Father, Vesulius, as is the custom among the Etruscans, so that from the records of the ancients I came to believe in the coming of the Great King, after whom the years shall be numbered. Therefore in my own Oracles I numbered the years as I knew from the auspices they would one day be numbered. But when my Father died prematurely, I came to Volterra because there was a College of Diviners in that City, but ever since Catilina betrayed his Fatherland, I have been assigned as Guardian of this Citadel by the Volterrans.[10]

Now Curzio Inghirami began to display some of the qualities for which he was habitually singled out among the young men of Volterra: a quick, nimble mind and an impressive depth of knowledge—his parents were certain that he had the makings of a great lawyer, and everyone else in town was inclined to agree. But Curzio's real love was history: he could therefore discern immediately from Prospero's mention of the traitor Catilina that the scrolls in both capsules must have been written around 62 B.C.

Lucius Sergius Catilina was a disgruntled Roman nobleman who lost the consular election in 64 B.C. to Cicero. A violent man of mesmerizing presence (and a very sore loser), Catilina then hatched a plot with a group of associates to seize the consulship he had failed to secure by vote. Before he and his fellow conspirators could set their plans in motion, Cicero denounced them before the Roman Senate in a series of four speeches, each recounting the progress of Cicero's own detective work amid outbursts of superbly elegant indignation, including the famous lament, "O tempora! O mores!"[11] As the consul unfurled his masterful oratory, Catilina fled north to

Etruria, where he had cultivated allies among the old Etruscan families by playing upon their resentment at Rome's policy of settling army veterans on former Etruscan lands. As always, too, the local slaves were ready to revolt in exchange for the promise of freedom. But Catilina had mistaken the caliber of his adversary. With the blessing of the Senate, Cicero, in full consular power, unleashed the Roman army on Catilina's ragtag forces and followed on his own initiative with a brutal purge of the plotters. By 63 the Catilinarian revolt was over. Catilina himself committed suicide before the Roman troops could arrest him and hand him over for Cicero, in the regalia of a triumphing general, to drag him in chains through the Roman Forum. Catilina or no Catilina, Cicero would celebrate the triumph anyway and published his Catilinarian orations for all Rome to savor.

Seventeen centuries later, crowding around Prospero the Etruscan's buried scrolls, Curzio and the other young Volterrans suddenly saw Catilina's conspiracy and its suppression with a piercing new immediacy. As schoolboys, they had all struggled through Cicero's four Catilinarian orations, unraveling the consul's intricate grammar while Cicero himself unfolded the story of Catilina's conspiracy. Mastering Latin as they watched Cicero master Rome, they could hardly have resisted crowing with the victorious Romans when the Fourth Catilinarian came to its close:

Vote, then, for your safety and that of the Roman people, for your wives and children, your altars and hearths, your shrines and temples, the houses and headquarters of the whole city, for sovereignty and liberty, for the safety of Italy, for the whole republic, vote carefully and strongly. You have a consul who does not hesitate to obey your decrees and, so long as he lives, shall defend your decrees and decisions and adopt them as his own.[12]

Now, abruptly, Prospero's scrolls turned the story inside out, telling it from the Etruscan viewpoint, and telling it in the very words, with the very handwriting, of an Etruscan eyewitness. Cicero's triumph, they could now see, had been Volterra's defeat, one more example of the Etruscans' relentless absorption by Rome.

In the beginning, the Etruscans had ruled the Romans, at least according to Roman tradition. But from the day when the people of Rome expelled their last Etruscan ruler, Tarquin the Proud, in 510 B.C., the little Latin republic had grown from a tiny city-state to a vast cosmopolitan empire. It began its expansion by gradually engulfing the other native peoples of the Italian peninsula, subjecting culture after culture to the dominion of the Latin language, Latin literature, and Latin customs. It took three hundred years for Rome to vanquish Etruria once and for all; the last dramatic episodes of the struggle were played out in the first century B.C., precisely in the time of Prospero the Etruscan, Catilina, and Cicero. By the time of the emperor Augustus (31 B.C.–14 A.D.), Etruscan writing had all but disappeared from tombs, buildings, coins, statues, books, everything except religious texts. As Curzio Inghirami and his friends read on and on through the cache of curled and yellowing scrolls, Prospero the Etruscan, waiting for Cicero's armies to close in on his citadel, bore vivid witness to the terrible price of Romanization:

When the Roman Army restored Fiesole and Volterra to Roman Dominion and laid siege to this Citadel, and I despaired of survival, I stored away my dear Household Gods and what money I had, together with the treasury of this Fort . . . and the Oracles written in Etruscan and Latin Letters. But because the Etruscan language has almost disappeared, I have summarized those [Oracles] that are in Etruscan letters. . . . I committed them all to the earth, so that they would not fall into the hands of the enemy, but if the fates permit it, may they be seen one day in a better light; otherwise let them be guarded here in perpetual eternity. . . .

I, Prospero of Fiesole, the Augur, in the year after Catilina.[13]

Clearly, as in the famous phrase of Tacitus, the Romans were once again to "create a desert and call it peace."[14] Prospero's scrolls made it clear that when Cicero eradicated Catilina and his allies, he had also purged Volterra of its largely Etruscan leadership. Prospero himself, the aspiring priest pressed into incongruous service as a soldier, gave tragic proof that many Etruscan men were already dead or under arms. Only thanks to a seer's foresight had Prospero consigned his testament to these strange capsules as his own culture faced extinction. The lonely hill of Scornello had been his refuge as it was Curzio's, with views across rolling fields toward Volterra on one side, and down chalky crags toward the river Cecina on the other. But Curzio could look out and daydream; Prospero spent his days in the grim search for Roman armies moving across the land.

As Curzio read on, Prospero the guardian's collected papers revealed the true extent of Etruscan priestly wisdom, a tradition about which both the ancient Greeks and Romans had spoken admiringly. To judge from the ancient scrolls, the augurs of Etruria seemed to have had more explicit foreknowledge of Christ's birth than the very Hebrew prophets, at least to judge from Prospero's references to "the prophesied King of the Jews." Furthermore, unlike the Hebrews, they had anticipated the Messiah's effect on the calendar: the prophesied King was also the figure "after whom the years shall be numbered."

The Etruscan augur's apparent foreknowledge of the Messiah would not have seemed strange in itself to young Inghirami; he and most other learned readers of his day had yet to subject ancient writings to detached anthropological scrutiny. Instead, these readers' relationship with the authors of antiquity was at once more personal and more intense. They looked to ancient religious texts for evidence that God had revealed His plan for human salvation long before putting it into effect in the early years of the Roman Empire with the coming of Jesus Christ. Furthermore, through the various traditions of ancient wisdom, from the monotheism of the Jews to the philosophy of the Greeks, they believed that God had carefully and gradually prepared the world to grasp the significance of Jesus' preaching, healing, death, and resurrection when these finally came to pass, at a specific time and place, in the Roman province of Judaea under the procurator Pontius Pilate.[15] By the same divine providence, the Lord had already set up the Roman Empire in order to disseminate the Gospel message widely and quickly once it was there to be proclaimed.

Out of respect for this long period of universal preparation, many Catholic scholars, at least from the fifteenth century onward, routinely bolstered theological arguments by referring them not only to Scripture but also to the works of the ancient Greeks and Romans. Church fathers and Christian apologists alike detected awareness of the Messiah's future advent in such disparate sources as Plato's philosophy, Virgil's poetry, and Homer's epics. They identified an earlier stage of this same preparatory period in the wisdom of the world's oldest civilizations: Persia, Egypt, and Babylon, calling these most venerable of all traditions the "primeval theology," *prisca theologia*. Classical authors themselves openly acknowledged that Egyptian wise men, Persian magi, and Chaldaean astronomers, among others, had shared their wisdom with the sages of classical antiquity, and to these venerable civilizations the Greeks and Romans always recognized their own profound indebtedness. It was no accident that the first people in the Gospel to acknowledge the birth of Christ were not just the humble shepherds, but also Wise Men from the East.

Italian scholars, especially, counted the Etruscans as ancient contributors to *prisca theologia* alongside the Egyptians, Persians, Babylonians, and Chaldaeans, for in antiquity both Greek and Latin authors traced many of Rome's social institutions and religious beliefs back to origins in Etruria.[16] In Tuscany, above all, the land of the Etruscans, the idea that ancient Etruria had created all that was civilized about Roman civilization was taken as compelling historical fact. This sense of communion with wise ancestors was still more pervasive in places like Volterra that retained imposing Etruscan ruins and a population largely descended from Etruscan stock. The discovery of Prospero's texts could finally bolster that vague fellow-feeling with concrete proof of the city's age-old importance.

Once Curzio's researches in Volterra had been crowned by evident success, the young antiquarian sent a courier off to his father in Florence, bursting with excitement about the new capsule and its documents.[17] Inghiramo Inghirami was staying as usual with a cousin, Cavaliere Giulio (1591–1639), former secretary to the Dowager Grand Duchess Christina, head of the granducal secretariat, and soon to become postmaster general for the Grand Duchy of Tuscany. The arrival of a courier, an expensive event in itself, implied extraordinary news, which Cavaliere Giulio heard as eagerly as Curzio's father. A man of more finely tuned political instincts than his country cousin, the head secretary insisted that a discovery of such consequence must be announced at once to Grand Duke Ferdinando. Working first through a high official (the prefect of the chamber, Orso Pannochieschi) and then through personal communication with the grand duke in person, Cavaliere Giulio convinced the young ruler to order further excavations at Scornello, excavations as thorough as a government-sponsored study could make them.[18]

On 29 December, therefore, Curzio assembled his largest expedition yet and headed out once again to the site. The workmen soon exposed a course of thick masonry walls, presumably part of what Prospero had called "the Citadel." In addition, Curzio reported, the workers brought up "fragments of marble, rusted iron, human bones, decayed and burned."[19] Amid the tangled roots of ancient oaks and ilex, they also found more capsules, but nestled close together, and so overgrown by vegetation that they could be pulled free only with difficulty. When cracked open, these, too, yielded new paper scrolls in Latin and Etruscan. Thanks to one of the new documents, the odd little bundles could be given a name at last: "scarith," a word that seemed to have no distinct plural form (fig. 7). There was also new information about Prospero himself:

Prospero of Fiesole, to the friend who shall discover these:

My father . . . taught me not only the Etruscan, but also the Greek and Hebrew tongues, and later the art of Augury and the secrets of nature herself, who provides all things for man. . . . [Yet] it is not prophecy that compels man, but the Great Aesar, who, when he had created man, created him as the possessor of his own will. . . .

I saw the three stars, Caris, Mor, and Turg in conjunction, and I saw from lightning on the plains of Asgaria, that what all the Scarith contain shall not be found, unless you, your friend,[20] and your father will have been present; know nonetheless, that if you, your friend, and your father [text missing] those stars shall be thrice evil for you.[21]

The meaning of "scarith" was evident from the context; this was (and is) the most common way for scholars to discover the meaning of Etruscan words. Because female tombs are consistently inscribed with the words *śec* and *puia*, whereas male tombs say *clan*, scholars feel confident in accepting that *šec* must mean "daughter," *puia* "wife," and *clan* "son" (no Etruscan man has yet been known to boast that he was someone's husband). In Curzio's time, however, none of these words had been translated; "scarith" was apparently the first. What Etruscan words he and his contemporaries knew—about twenty in all—had been preserved in ancient Greek and Roman literature, and one of these words was "Aesar." According to the ancient authors, Aesar meant "God."[22]

But the discolored scroll did more than confirm the meaning of Aesar. Prospero's claim that his "Great Aesar" had created humanity as the possessor of free will showed a striking parallel between Etruscan religion and Catholic doctrine of the seventeenth century.

Exemplum Scarith, in quo sequentes asseruabantur Scripturæ.

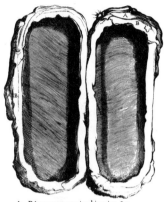

A. Primus cortex ex pice, bitumine, &c.
B. Secundus cortex ex solidiori materia.

7 : Cutaway view of a scarith, copperplate engraving. Curzio Inghirami, *Ethruscarum Antiquitatum Fragmenta* (Florence: Massi, 1636), 6. Photo: Antonio Ortolan.

One of the most divisive controversies to split seventeenth-century Christendom involved the choice between Calvin's doctrine of predestination and the Catholic position that salvation was justified by good works as well as faith. On the basis of this and other doctrinal disagreements (not to mention a good many festering political disputes), the Thirty Years' War had raged for most of Curzio's life.[23] Now Prospero of Fiesole provided direct evidence from *prisca theologia* that the Catholic position had always been the correct one.

Prospero's mention of the stars Caris, Mor, and Turg confirmed that the Etruscans had also been expert in astronomy: this was another surprising revelation, and potentially an explosive one—only the year before, on 22 June 1633, Tuscany's greatest astronomer, Galileo Galilei, had been sentenced by the Roman Inquisition to perpetual imprisonment, commuted in late 1634, as the scarith emerged from the ground, to house arrest in the suburbs of Florence. Galileo had been forced in addition to "abjure, curse, and detest . . . a doctrine which is false and contrary to the divine and Holy Scripture: that the sun is the center of the world and does not move from east to west, and the earth moves and is not the center of the world."[24] Yet if Galileo himself, old and nearly blind, was now silent on matters of cosmology, his new invention, the telescope, had opened up a cosmos thronged with new stars, planets, and moons. The three Etruscan stars Caris, Mor, and Turg cried out for identification, and one of Galileo's own students came to the rescue, Father Vincenzo Renieri, who set to work shortly afterward on interpreting the astronomical lore contained in the scarith texts.[25]

On a less advanced plane of science, Prospero's predictions from "lightning on the plains of Asgaria" provided dramatic proof that Etruscan thunder-divination had been as important to Etruscan thought as the ancient authors suggested. The summer skies of Tuscany, raked by periodic thunderstorms, had always supplied ample material on which to practice the characteristic art of a priesthood that the Romans called the *fulguriatores*.[26] Late Latin authors made it clear that Etruscan books on the interpretation of thunder were in constant use in Rome until the end of the sixth century A.D.—more than two centuries into the Christian era. Yet the Nordic ring of "Asgaria," with its evocations of Thor hammering thunderbolts on the plain of Asgard, also struck a chord of its own. The similarity between Scandinavian runes and Etruscan script had already been noted for a century among Europe's most learned circles, beginning with the Swedish Catholic scholar Johannes Magnus in the mid-sixteenth century.[27] However, with the increasing ascendancy of Scandinavian commercial and military power in the seventeenth century, these parallels took on an added diplomatic significance by helping to forge a cultural bridge between Mediterranean Catholic south and Baltic Lutheran north. In 1597 the Belgian scholar Bonaventura Vulcanius had greatly facilitated research on such cultural connections with his study on northern tongues, *De literis et lingua Getarum*, a book that included some of the first transcriptions of runes.[28] Once again, it seemed, Prospero's texts had emerged from oblivion at a singularly auspicious moment.[29]

Unfortunately, however, the fate of Prospero himself looked increasingly dire as further excavation brought forth more scarith. In January 1635 Curzio came across a scarith with its outer layer scorched away. Nearby, a battered tin statue, the top of its head missing and Etruscan letters on the hem of its skirt, lay in a tangled heap as if it had been flung to the ground and left there to molder (fig. 8). In that pitiful figurine, Curzio recognized the sole survivor of Prospero's "dear household gods." A broken bronze lamp (fig. 9) completed the site's picture of sud-

8 : Tin "Etruscan" household god, copper-plate engraving. Curzio Inghirami, *Ethruscarum Antiquitatum Fragmenta* (Florence: Massi, 1636), insert. Photo: Antonio Ortolan.

9 : Fragment of Etruscan bronze fibula, seventh century B.C.E., labeled "lamp," copper-plate engraving. Curzio Inghirami, *Ethruscarum Antiquitatum Fragmenta* (Florence: Massi, 1636), insert.

den devastation long ago. "We arrived at the conclusion," Curzio wrote, "that when the Romans captured Scornello, they carried [Prospero] off and then toppled the Citadel to the ground."[30]

The Investigation

1635

The Vulture and the Snake shall be joined together, and Doves shall be born of them; these shall fly like Eagles, but they shall not reach their goal. They shall await the dew, and there shall be no dew, neither shall the rains slake their thirst; truly they shall taste the Tyrrhenian Sea. SCARITH #109, DISCOVERED 7 MARCH 1637[1]

For a seventeenth-century scholar, the proper place to report an exciting discovery was a learned society. These came in all shapes, sizes, and nationalities, from the tiny but influential Lincei in Rome, to whom Galileo had first reported his findings with the telescope, to state-sponsored bodies like the Oziosi of Naples, founded in 1611 under the auspices of the Spanish viceroy, or the Florentine Accademia della Crusca, dedicated to the promotion and purification of the Tuscan dialect.[2]

Curzio Inghirami and Father Vadorini therefore turned their treasures over to the scrutiny of Volterra's leading learned academy, the Sepolti, or "Buried Men," "buried" because their great love of knowledge made them as good as dead and buried

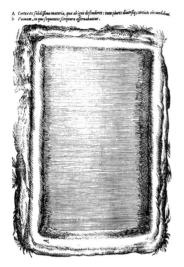

A *Cortex ex solidissima materia, que ab igne defenderet : tum plures diversisq; corticis circumdabant.*
b *Vacuum , in quo sequentes scriptura asservabantur .*

10 : Cutaway view of Scarith #6, woodcut. Curzio Inghirami, *Ethruscarum Antiquitatum Fragmenta* (Florence: Massi, 1636), 92. Photo: Antonio Ortolan.

to the ephemeral world.[3] Dead to the world the Sepolti may have been, but they were certainly alive to argument. Riffling the scarith's discolored pages (fig. 10), the Buried Men began to air their doubts. Ancient writers like Livy and Pliny, some declared, reported without exception that the ancient Etruscans wrote their books on bolts of linen, not on paper. From the thunder-divining *fulguriatores*, to the liver-examiners (*haruspices*), to the bird-diviners known as augurs, Etruscan priests, they claimed, had all committed their sacred lore to the sturdy fibers of finely woven flax. They were right: one of these Etruscan linen books has survived to the present day, reused to wrap an Egyptian mummy and preserved in the dry air of the Nile Delta. However, this remarkable artifact was not identified until the twentieth century; neither were the many representations of linen books on Volterran cinerary urns, which simply look like — and are — folded cloth.[4] The Sepolti reached their conclusions from literature alone.

Curzio's reply to these objections, swift, elegant, and erudite, gave a glimpse of the lawyer he might one day become: the paper scrolls of Scornello, he asserted with newfound authority, now showed that the "linen books" mentioned by the ancients must have been written on paper made of linen rags, just like the paper of the seventeenth century. In the absence of any Etruscan cloth books to prove the contrary, the scarith pleaded a tangible case;

Curzio's argument looked all the more impressive because he provided archaeological as well as literary evidence to back it.

In Florence, meanwhile, Grand Duke Ferdinando kept track of the excavations' progress through Curzio's influential uncle, Cavaliere Giulio. Only four years older than Curzio himself and intellectually inclined, Ferdinando de' Medici must have been particularly struck by the idea of a twenty-year-old holding his own so gracefully in scholarly debate. He summoned Curzio to court for a private showing of the scarith, followed by public presentations before the professors of the two granducal universities, Pisa and Florence. Disarmingly, Curzio recounted subsequent events from the scarith's point of view: "They were taken to Pisa, where the most learned scholars of its Gymnasium were summoned to inspect them. The opinions were divergent; some recognized their antiquity and proclaimed it, while some, who suspected a malicious hoax, asserted that linen paper had not existed among the Etruscans."[5]

Undaunted, Curzio scooped up scarith, tin statue, and battered lamp:

I sought out Florence and I put them on display for the Grand Duke, the Most Serene Princes, the whole University, and the most scholarly men. Some, admiring their antiquity, were struck by the material of the Scarith, which protected against water on the outside and fire on the inside, but some, such is the variety of human temperaments, declared that they were factitious, and forged.[6]

At this point, Curzio reported, the grand duke decided to call in the police. Inghirami himself claimed that their investigation aimed at "settling the controversy among scholars," but it is also clear that the investigators had another mandate: to look out for evidence of fraud.

The police in Curzio's day were a body constituted not in Volterra itself but in Florence, to which Volterra had owed allegiance as a vassal state ever since the brutal siege by Lorenzo de' Medici in 1472.[7] In that same year, one that Volterrans still recall with undimmed resentment, the victorious Florentines set a grim fortress atop Volterra's citadel (fig. 11), a modern structure designed to repel guns as well as catapults. (It now serves as a maximum-security prison.) From within this ominous reminder of Florentine dominion, the Provveditore (Overseer) of Volterra, Tommaso de' Medici, commanded his squadron of police in the 1630s.

However, Florence in the seventeenth century differed considerably from the Florence that had conquered Volterra in 1472. Still governed by the Medici (although by a different branch), the Florentine state had gradually engulfed Siena, Pisa, Livorno, and Arezzo to create a political unit known as the Grand Duchy of Tuscany. No longer a nominal republic (as it had been for its fifteenth-century Medici rulers, Cosimo the Elder and Lorenzo the Magnificent), the Grand Duchy was a full-fledged European monarchy with a ruling dynasty; an extensive bureaucratic apparatus; ties of politics, commerce, and marriage throughout the Continent; and summer capitals in the port cities of Pisa and Livorno.[8]

The police were well aware that within this larger Tuscany, as in Volterra itself, Curzio Inghirami belonged to a family of consequence. Indeed, the Inghirami fortunes had reached new heights in the early seventeenth century, when Curzio's great-uncle Jacopo Inghirami was named admiral of the Fleet of Saint Stephen, the Tuscan navy.[9] Founded in 1561 to protect the Tuscan coast from the menace of "Saracen" pirates (most of them Corsicans and Turks), the Fleet of Saint Stephen was governed by a brand-new order of maritime knights, dedicated to Saint Stephen and modeled on the old crusading order of the Knights of Malta.[10] Along with Admiral Jacopo, eleven other Inghirami men had been elevated to these naval Knighthoods

11 : View of Volterra from Villa Inghirami, Scornello. Author's photo.

of Saint Stephen, including Curzio's grandfather (another Curzio) and no fewer than ten uncles, Cavaliere Giulio the most authoritative among them. Curzio himself would eventually lay claim to the same knightly rank.

In a society that put great emphasis on fine distinctions of status, the Inghirami felt no hesitation about flaunting their family prestige. In 1607 Admiral Jacopo paid the immense sum of thirty thousand scudi to turn the left transept of Volterra's cathedral into a private chapel for the Inghirami family, replete with marble altar, altarpieces by famous painters like Domeni-

27

chino, and walls frescoed by another prominent artist, Giovanni da San Giovanni. Executed at the time of Curzio's birth, these paintings offer a suggestive portrait of the family and the social expectations that shaped the young amateur archaeologist as he was growing up.[11]

The most striking of Giovanni da San Giovanni's frescoes shows a newly converted Saint Paul heaving into Damascus, filled with the Gospel and eager to preach it. Damascus has been outfitted for the occasion with a twin of Volterra's Romanesque baptistery, emphasizing the universal applicability of this dramatic Bible story to the profession of apostolic faith; it says, in essence, that to the committed Christian, every town, from Volterra to Beijing, is a new Damascus in need of the Gospel.[12] Off to the right side of the painting, an assembly of Inghirami elders looks upon the dramatic scene with an aristocratic languor that belies the fresco's eager evangelical message. The chapel's benefactor, white-haired Admiral Jacopo, Marquess of Montegiovi, stands to the far right, looking out at the spectator. Beneath him, in priestly robes, his brother Antonio, canon of the cathedral, watches Saint Paul intently. Several other figures in contemporary dress share the broad faces and jovial expressions common to most of the Inghirami family; they must include some of the admiral's many nephews, probably including Cavaliere Giulio and his brothers Cavaliere Giovanni and Cavaliere Tommaso Fedra, all of them Knights of Saint Stephen in their own right. Clad in the lace ruffs and black silks of early seventeenth-century grandees, honorary medals well in evidence, the Inghirami clan poses with the studied ease of Olympian gods. Twenty years later, they must have faced the grand duke's police investigators with an equally impressive display of ease and amusement. Objective inquiry into young Curzio's discovery cannot have been an easy assignment.

But the investigation could not have been any easier for the Inghirami themselves. Every adult connected with the scarith knew that the controversy over their authenticity had already gone far enough to bear on the rest of Curzio's future and that of his sister, Lucrezia.[13] Their original discovery may have been a matter of chance, but once the objects had become public, they affected the family's prestige, all the more so because Curzio belonged to a less distinguished branch of his prominent clan, isolated on their property in the hills above the salt springs of Le Moie. Hence his success at presenting his antiquarian finds would also determine his success at presenting himself. If the scarith themselves were subject to discussion, so was Curzio's reputation and so, in turn, was the family's standing in Tuscan society. The scarith promised their discoverer and his family either fame or shame, but probably not both.

In facing the investigation, Curzio could reasonably expect immunity from only one charge in a police investigation that could only compound the potential risks to his reputation: the charge of fraud. However vocally skeptics in Volterra, Pisa, and Florence might have voiced their doubts about the scarith's authenticity, none was yet inclined to accuse a nobleman, much less one so young, of mounting the deception himself. Instead, they suggested, his love of antiquarian study must have exposed him to the wiles of some other deceiver. But every other speculation about the scarith and their maker flew freely as the police went about their inquiries.

Old salts and landed gentry, accustomed to courtly intrigue and to command, the elder Inghirami closed ranks around their vulnerable young charge. Curzio quickly discovered that, for all practical purposes, his life had been turned over to a triumvirate of male relatives: his father, Inghiramo, his uncle Cavaliere

Giulio, and his cousin Cavaliere Francesco, the latter two Knights of Saint Stephen who boasted firm ties with Grand Duke Ferdinando. But Cavaliere Giulio, the eldest and most distinguished, was clearly in charge.[14]

The police, in turn, had their own agenda to pursue. Archaeological finds, like cultural matters in general, played a role of unusual significance in the Tuscan Grand Duchy. An uneasy and somewhat artificial alliance of former republican city-states (several of these torn by centuries of internal factional strife as well as traditions of stubborn independence), Medicean Tuscany lived in constant fear that its various components would splinter apart. The man who first cobbled it all together, Grand Duke Cosimo I, had resorted to a variety of expedients to keep his state united, an extensive bureaucracy not least among them. A trio of capital cities kept the grand duke moving among his subjects, patiently securing the restive port of Pisa while he built up Livorno as a second port of recourse should Pisa ever secede.[15] Harping on a genuine Saracen menace (a Sienese girl who was famously kidnapped from the Tuscan coast in the sixteenth century ended up as the favorite queen in the sultan's seraglio), the Fleet and Knights of Saint Stephen bound Tuscany's two port cities and all Tuscans together against a common enemy, all the more effectively because that enemy, Ottoman Turkey, was an infidel state.[16]

In many ways, however, what best united the Grand Duchy was the glory of its cultural institutions. The universities of Florence and Pisa attracted faculties of international distinction. The ducal galleries of the Uffizi testified to Tuscany's past greatness and contemporary renaissance with superb collections of antiquities and modern art. A significant assemblage of Etruscan artifacts in the grand duke's collection suggested that a distinctively Tuscan affinity for the fine arts had existed from time immemorial.[17] Curzio and his family could hope for no more illustrious destination than the grand duke's Etruscan museum for at least some of the scarith. More peculiarly to Florence, the powerful Accademia della Crusca, Tuscany's most illustrious and most exclusive gentlemen's academy, had taken up the charge of "purifying" Tuscan vernacular, on the one hand, by bringing it back to what the academy regarded as its original Etruscan vocabulary, and at the same time elevating it to the status of a full-fledged European diplomatic language by using it for official correspondence.[18]

As the sixteenth century turned into the seventeenth, and the Grand Duchy exerted less and less economic and political power on the outside world, the importance of culture to the Medici state became ever more central. The myths of earlier Medici—from the fifteenth-century banker-aristocrats, Cosimo the Elder and Lorenzo the Magnificent, to the sixteenth-century warrior Cosimo I, founder of the granducal dynasty—maintained Florentine pride of place as a civilizing force in a continent that had elsewhere fallen prey to the bitter religious strife of the Thirty Years' War. Despite his youth, Ferdinando II was already fostering art, writing, and learned research with great enthusiasm, abetted by his acutely intellectual brother, the future cardinal Leopoldo.[19] This was the environment into which the scarith and their discoverer, under the firm control of Cavaliere Giulio, stepped when they met the Volterran police.

The police force and its chief, the Most Serene Provveditore Tommaso de' Medici, were surely prepared to defer to the Inghirami as a matter of prudence. But Cavaliere Giulio, conscientious paterfamilias of the vast Inghirami clan, was taking no chances. In late March 1635, he wrote a long report to Lord Bailiff Andrea Cioli, secretary of state of the Grand Duchy of

Tuscany and hence also Inghirami's longtime colleague. With a professional precision, he set out the circumstances of Curzio's discoveries and countered the chief objections to the scarith's authenticity:

This evening the Signori Inghiramo and Curzio Inghirami, the former my Cousin and the latter my Nephew, went to City Hall [in Volterra] to show the Most Serene Provveditore [Tommaso de' Medici] the rare monuments found by Them in a Villa that my Father had exchanged with them. . . . There are many papers written with Tuscan Characters, or better ancient Etruscan, and although there has been debate among scholars about whether our Paper existed in those times, experience teaches us that the Linen Paper mentioned by Livy and Pliny is the kind that we use today. . . .[20]

With the same concision, Cavaliere Giulio also summarized his view of the documents' significance:

These begin with Noah founder of Volterra and contain a continuous series over 1800 years of 55 Tuscan Kings, the foundation of the 12 Cities [of the Etruscan League]. . . . There are also predictions of the advent, life, and death of Our Lord Jesus Christ, clearer than the ones preserved in the Bible, among them one that said: "A Great King is coming after whom the years shall be numbered."[21]

Noah's role in the founding of Volterra was a new revelation of the scarith, but Noah's connection with Etruria was an old story to seventeenth-century Italians.[22] A twelfth-century guidebook to Rome, the *Graphia aurea urbis Romae*, had already identified the Hebrew patriarch with the ancient Italic god Janus, reporting that once the Ark had landed safely on Mount Ararat, Noah had taken to traveling the world by raft, vigorously repopulating the flood-drenched continents with the heroic aid of his wife. He had traveled up the Tiber as far as

Rome, where he took a new name, Janus, in honor of his introduction to that region of wine (Hebrew *yayin*), and applied the new title to his Italian place of residence, the Janiculum Hill. In the late fifteenth century, this legendary identification of Noah with Janus gathered further momentum, especially after a Dominican friar named Annius of Viterbo began to circulate a set of spurious ancient texts that repeated the story of Noah, offering these "discoveries" as proof that the Etruscan language descended from prophetic Hebrew.[23] Long after Annius and his texts had been largely discredited on historical, linguistic, and theological grounds in the wake of the Reformation (Martin Luther used them to bolster his theses), the old association of Noah and Janus persisted in Italy, as it had already for at least four centuries and may have for many more previously.

The old tradition of Noah's presumed presence on Italian soil lent a certain plausibility, at least in the eyes of Curzio's contemporaries, to Prospero the Etruscan's foreknowledge of Christ; if anything, the young augur's prophecies offered concrete proof that Etruria had enjoyed special access to Hebrew revelation ever since the age of the patriarchs. If Etruscan priests had handed down Hebrew wisdom directly from Noah, then it only made sense that Prospero would have some foreknowledge of Christ's advent and of the future Christian calendar. Certainly Volterra's rugged Etruscan walls reinforced the impression that the city's ancient history must reach back as far as patriarchal times.

Cavaliere Giulio, however, had no intention of waiting passively for the police to pass their judgment on the scarith's authenticity. His letter to Andrea Cioli shows that he faced the approaching inquiry with his own set of well-defined intentions for its various stages. Certifying the scarith's antiquity was only the first step in his plan; more urgently, he wanted the police chief, the Provveditore, to award his family publication rights to what people were already beginning to call "the Etruscan Antiquities":

I believe that the Most Serene Provveditore will allow us to publish them because they are objects found in our Villa, in the possession of Our Household for more than 450 years in succession without our having known from any of our Elders that in ancient times there had existed a Villa, or a Fortress, nor would it have been known had chance not brought it about that the aforementioned Signor Curzio and the Signorina Lucrezia his sister found a bundle of pitch, hair, and lime, within which there was a Paper saying "You have discovered the treasure. Mark the spot, and go away."[24]

Publication in seventeenth-century Tuscany was no simple matter. With the Inquisition in full flower, manuscripts set for publication in Catholic cities first had to pass scrutiny by a battery of ecclesiastical censors, who were managed by the local bishop and ultimately responsible to the Holy Office in Rome. Just to be safe, Catholic governments like the Grand Duchy often maintained their own separate set of civic censors, who answered to the city administration but served the same purpose of scouring written work for hints of "matter injurious to doctrine or morals." Theological books could not go to press in Catholic cities without bearing the ecclesiastical censor's coveted authorization, "let it be printed" — "Imprimatur" — and the bishop's confirmation of that order. (In Tuscany, for patriotic reasons, the order to print was often emitted in the "Etruscan" vernacular, "si stampi.")

After publication, of course, a book might still fall victim to charges of heresy; a pitfall the civic censors were designed to head off. To root out doctrinal error among books in print, the Inquisition in Rome maintained its own battery of censors who investigated reports of doubtful publications and singled out the most pernicious of them for listing in their dread Index of Prohibited Books. This is what happened to Galileo's *Dialogue on the Two Chief World Systems*, published in 1632 and indexed the year after. Indexed books could not be imported into Catholic states without a special license, and faithful subjects could only read them by requesting special permission from Rome. The effects of this pervasive censorship on the Italian printing industry were dire. Civic authorities in seventeenth-century Venice and Naples were still resisting church interference in their publishing industries, and despite occasional skirmishes with the Inquisition, their presses continued to work actively. In Florence, on the other hand, the printing industry had shrunk drastically from its glorious beginnings in the fifteenth and sixteenth centuries.

To make matters worse, the scarith texts, because of their prophecies and their astronomical lore, could not be considered wholly safe from inquisitorial questioning. Hence Cavaliere Giulio, in a second letter to Andrea Cioli, made clear his intent to seek a formal writ of authentication for the scarith from the papal nuncio to Florence as well as the grand duke's official approval. Cavaliere Giulio's second letter was written only two days after its predecessor, showing that in fewer than forty-eight hours he had already convinced the police to grant him full rights to publish the scarith. So long as he had permission to publish, moreover, he seemed to have believed that authentication could wait:

This evening was the last session about the texts of Volterra, and finally among these Signori the Critics it is concluded that they are the most noble Manuscript Texts, to our knowledge, that have so far been found in Europe. The Most Serene Provveditore is content that they should go to press, and we shall hold a Trial in Volterra about the manner, time, and by whom they were found, and officially recognize [their authenticity], and we shall have the same done in Florence, and before the Tribunal, and before the Monsignor [Apostolic] Nuncio, and in Copper Plate, as large as the pieces of paper on which they have been written, so that this Script will be imitated as closely as possible, so that the World may enjoy so noble a treasure, which ought to be welcome to Italy because here is her origin from the Flood onward.[25]

Publication, however, would require more than church approval; it also required money. Most seventeenth-century books begin with an elaborate preface to the patron who had generously underwritten the expensive process of printing. Cavaliere Giulio's letters to Andrea Cioli show that he envisioned publishing the scarith in as lavish a format as possible, in folio edition, illustrated with copper-plate engravings, all testimony, as he insisted, to the scarith's remarkable importance. In turn, however, Curzio Inghirami would no longer be free to spend his time as a carefree afternoon angler; he would have to put his proliferating scarith texts into some coherent order and prepare a manuscript for press. As much as the scarith themselves, he had become a cultural property, with his scheming uncle evidently in charge of both.[26]

With publication rights in hand, the Inghirami pressed the Provveditore for a decree of authenticity, only to discover that whereas negotiation for a license to print took only a day or so, the procedures for authentication could grind on with exquisite slowness. It was June 1635 before Provveditore Tommaso de' Medici made his first official visit to the site at Scornello, accompanied by Ottavio Capponi, supervisor of the Salt Tax for the Grand Duchy.[27] For the next two weeks, the two officials alternated days in the field with days spent examining witnesses, and there is no doubt which days they enjoyed more. Proud of their own Etruscan heritage, they found the dig at Scornello an intoxicating immersion in their ancient past.[28] Thirteen new scarith emerged from the chalky soil during their seven days of on-site investigation, two of the capsules wrested from a gnarled tangle of oak tree roots by none other than Ottavio Capponi himself. Their direct experience of amateur archaeology had a forceful effect on the granducal agents: after a trip to view the oldest documents in Volterra's civic archive, they officially declared that the scarith of Scornello had been genuinely buried and were very, very old. With professional circumspection, however, they let their verdict fall short of declaring the scarith texts Etruscan; the two bureaucrats openly (and prudently) acknowledged their own limits as antiquarians.

By now, the grand duke had become sufficiently curious to believe that another on-site visit to Scornello might be of use; Curzio himself recalled that "[his] Highness . . . sent Messrs. Mario Guiducci and Niccolò Arrigheti, Florentine Gentlemen, specifically to examine the site and observe the excavations, and their statement is still preserved in the Civic Archive of Volterra."[29]

Late in 1636 Curzio (urged along by Cavaliere Giulio) finally committed what he called "the Etruscan Antiquities" to press in Florence, despite the fact that censorship was not the only drawback to publishing in the capital city.[30] Seventeenth-century Florentine printers used paper from factories in nearby Colle di Val d'Elsa, all of which had been maintained by the Florentine government since the late sixteenth century as a means of financing the Monte di Pietà, the civic pawnbroker. It came as no surprise to anyone that in seventeenth-century Florence the price of books was high and the output low, or that quality usually succumbed to the temptations of low-grade paper, worn-out typefaces, and recycled illustrations.

Strikingly, however, Curzio Inghirami's *Ethruscarum Antiquitatum Fragmenta* suffered not one of the standard indignities to which Florentine books were subjected in these dark days of the seventeenth century. Crisply printed on expansive folio sheets of good paper, it boasted a host of illustrations, both woodcuts and expensive copper-plate engravings, sometimes, remarkably, on the same page (fig. 12). The printer of this outstanding example of his art was a new arrival in Florence, Amadore Massi from Forlì in Umbria, who had just set up shop for the first time in 1636.

Yet despite the fact that the Inghirami had sought official permission from Florentine authorities to proceed with publication, Massi issued his book under a false imprint and a false date: the title page of *Ethruscarum Antiquitatum Fragmenta* (fig. 13) reads Frankfurt 1637, though the book's printing was actually completed in Florence in 1636. Neither does a printer's name appear, but only a pair of Etruscan letters flanking a caduceus.[31]

Often such elaborate secrecy, a common practice in the seventeenth century, was aimed at dodging the censors; if Protestant cities offered authors a refuge from Catholic censorship, the spurious imprint of a Protestant city might furnish some of the same protection, in addition to giving Curzio's controversial Etruscan Antiquities a touch of cosmopolitan cachet. The Frankfurt imprint may also have disguised the extent to which the grand duke himself was personally involved in bringing about publication.

In any case, there is no doubt that in a time of economic hardship considerable sums of money went into printing the *Ethruscarum Antiquitatum Fragmenta*, presumably supplied in large part by Cavaliere Giulio, who, like Admiral Jacopo Inghirami almost thirty years before, cultivated a large extended family as an expression of his authority in the world.[32] Furthermore, Giulio Inghirami harbored a personal interest in Scornello, which had originally belonged to his father.[33] Grand Duke Ferdinando may also have supplied a subvention.

Caue, Caue, Caue.
Profperus Fefulanus huius Caftri accola, arcis Cuftos Vaticinatus eft anno
poft Catilinam extinctum.
Thefaurum, inueniſti locum figna, & abi.

Anno à rege iudæorum nunc†dæcm D C xxiiii crucifix o M D L xxxxi
Fi ádes ciuj uitam inuenient non ouident ſed dormient

Ueniet canis fideliter ſeruiet ſeruitute libera. IX annos et amplius
lupa mater agni . agnus amabit canem . Ueniet porcus de grege porcorum
et deuorabit labores canis

Caue Caue Caue

A

Profperus Fefulanus huius caſtri accola. arcis cuſtos uaticina
tus eſt anno poſt Catilinam extinctum
Thefaurum inuenni ſti . locum ſigna . & abi

Perlegi, & obſtupui ; locum ſignaui. Domum reuerſus Inghiramo
Patri, quæ acciderint, enarro ; ipſe mecum & ſchedulam, & notas ri-
matur : tandem vterque, cum de theſauro agi intelleximus, locum effo-
diendum decreuimus : ſequenti die effoditur, fictilis cuiuſdam ſolum-
modo vaſis fruſta offendimus ; hinc ſuſpicantes aut Theſaurum ab alijs
ſubreptum, aut nos illuſos ab aliquo fuiſſe, opus dimiſimus. At cum anni tempus rure frui permitteret, Dominicus Vadorinius Sacerdos, & fa-
miliaris noſter ad ſacrum faciendum Feſto die S. Luciæ aduenit, cui, &
chartam inuentam oſtendo, & vanam de Theſauro noſtrâ ſpem ſubri-
dens enuncio. Ipſe pauliſper meditatus Vaticinium, deſidiam incre-
pans meam, ad opus reſumendum hortatur. Accitis itaq, colonis ad ef-
fodiendum redeo, quo loco primum Scarith reperi (hoc ſiquidem nomine huiuſmodi inuolucra à Proſpero appellantur) dum altius ſolum fodi-
tur, fundamenta validioris, & craſſioris parietis offenduntur, & in his
vrnula quædam lapidea ; quæ effracta, dedit ſecundum Scarith, characteribus, quos cernis, Ethruſcis antiquitate corroſis inſignitum.

B

C

Hoc

12 : The first scarith text, copper-plate engraved image, woodcut frame, and movable type. Curzio Inghirami, *Ethruscarum Antiquitatum Fragmenta* (Florence: Massi,
1636), †ii 3r. Photo: Antonio Ortolan.

D cl. II. 10

ETHRVSCARVM
ANTIQVITATVM
FRAGMENTA.

Quibus Vrbis Romæ, aliarumque gentium
primordia, mores, & res geftæ
indicantur

A

CVRTIO INGHIRAMIO
REPERTA
Scornelli propè Vulterram.

DVPLEX INDEX
omnia edocet.

FRANCOFVRTI
Anno Salutis M. DC. XXXVII.
Ethrufco verò cIɔ cIɔ cIɔ cIɔ CCCCXCV.

13 : Title page of Curzio Inghirami's *Ethruscarum Antiquitatum Fragmenta* (Florence: Massi, 1636). Photo: Antonio Ortolan.

Whatever their source, dozens of scudi ensured that Curzio's antiquarian study could speed into production, without time spent seeking out a patron or composing a laudatory preface; instead, the book's ancient texts, accompanied only by Curzio's brief description of the initial find, stood splendidly on their own. Two foldout maps, drawn for Curzio by Father Domenico Vadorini, show Volterra in its Etruscan glory days, a bustling city packed with temples, theaters, and obelisks, and the sadly reduced Volterra of 1636. "This is how much distant antiquity can change with time," mourns the caption of the second map (fig. 14). Vadorini also supplied a map of the site at Scornello, a panoramic view of the foundation walls of Prospero's bastion laid out across the Volterran hinterland (fig. 15). Throughout the book, copper-plate engravings reproduce many of the scarith texts at length, while others are rendered in full-page woodcuts. The different types of scarith casings are shown in actual state and in cutaway view so that readers who would never have a chance to handle a real scarith could garner a reasonable idea of what the capsules were like. The number of excavated scarith had mounted by this time to 109; helpfully, the book records each capsule's date of discovery in the margin. A more professional hand than Vadorini's engraved an image of Prospero's poor battered household god and the bronze "lamp." As final proof of the volume's serious intent, Curzio furnished his preface with a two-page foldout engraving of the Inghirami family tree, an oak with the shield of their grand ancestor, a tenth-century Saxon named Enno Billing of Lauenburg, propped at the base of its massive trunk. Far above, the banner of Curzio di Inghiramo flutters proudly among the mighty tree's topmost branches. The book, in short, was gorgeous. If John Keats was right to say that "beauty is truth; truth beauty," then the *Ethruscarum Antiquitatum Fragmenta* provided truth of the highest order.

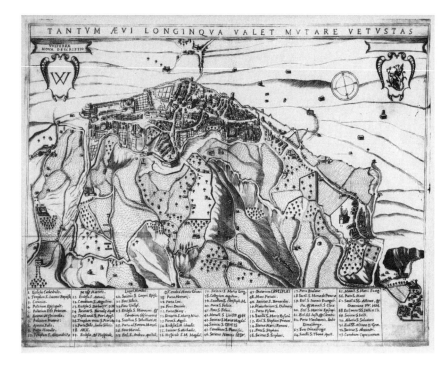

14 : Volterra in 1636: "This is how much distant antiquity can change with time"; Curzio Inghirami, *Ethruscarum Antiquitatum Fragmenta* (Florence: Massi, 1636), insert. Photo: Antonio Ortolan.

Like the book's false Frankfurt imprint, its considerable beauty had a clear purpose, for Curzio Inghirami would need powerful allies when his scarith entered the larger scholarly world. With the authenticity of his find still in question from many quarters at home, the *Ethruscarum Antiquitatum Fragmenta* carried Prospero's scrolls beyond the borders of Tuscany, making their attractive appeal to an international audience of antiquarians. At the same time, and perhaps deliberately, the book removed the burden of authentication from the physi-

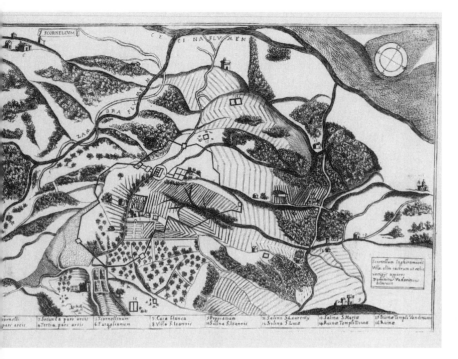

15 : The site of Scornello, as drawn by Father Domenico Vadorini, 1636. Curzio Inghirami, *Ethruscarum Antiquitatum Fragmenta* (Florence: Massi, 1636), †ii v–†ii 2r. Photo: Antonio Ortolan.

cal evidence of the scarith themselves to the quality of the texts they contained. In Volterra, Pisa, and Florence, the Inghirami had discovered that scholarly debate reverted incessantly to the problems raised by the scarith as physical objects, focusing with particular intensity on the fact that they were made of paper; no matter how smartly Curzio protested, some scholars continued to insist that Etruscan linen books had been made of real cloth, not rag paper. Meanwhile, Prospero's astronomy, calendrical lore, and Etruscan histories had received much less attention. Thanks to the printed *Ethruscarum Antiquitatum Fragmenta*, the arguments about the artifacts' authenticity could now be taken up by people who might never see a scarith and who might, therefore, judge Curzio Inghirami's discovery, and his ancient friend, Prospero of Fiesole, by what the scarith could offer to religion, astronomy, literature, and history as well as archaeology.

The Spy

1638

Μέγα βιβλίον, μέγα κακόν
Big book, big trouble.

CALLIMACHUS, PARAPHRASED

As Socrates observed in Plato's dialogue *Phaedrus*, writing can be a dangerous thing, for books fall into unpredictable hands. When *Ethruscarum Antiquitatum Fragmenta* moved the scarith and their controversies out beyond the borders of Tuscany, Curzio Inghirami faced precisely the problems that Plato had warned about. Designed to elude nagging objections to the Volterran Antiquities as physical objects, the book also raised new kinds of questions about the scarith and their discoverer.

Curzio Inghirami had built his reputation hitherto on his conduct in live debate, at Volterra, Pisa, and Florence. In those public demonstrations, he had impressed his hearers with his youth, quick wit, courtly manners, and conspicuous learning. True or false, the scarith and their defender provided excellent entertainment. Curzio's audience so far had also come from a relatively small, tightly knit social milieu made up of men attuned to each other's ornamental speech, instinctively inclined to analyze vocabulary and pronunciation for "Etruscan" purity, to note euphonic cadences and rolling rhetorical periods, to mark a man's standing by the way he spoke as well as by what he said. Among the gentlemen of Tuscany, the rules of debate, like the rules of evidence, adhered to a set of clear, predictable standards. They were the same standards that had governed Galileo Galilei's presentation of his scientific work; they also governed the intricate proceedings of the Accademia della Crusca as it carried out its mission to purify the "Etruscan" dialect.

Yet, as Galileo learned too late and Curzio was soon to discover, these rules of courtly conduct changed abruptly at the Tuscan border. Both of them may have assumed that Rome, under the Florentine pope Urban VIII, was simply another Florence farther south. But it was not.

Pope Urban VIII may once have been a Florentine gentleman named Maffeo Barberini, but after his elevation to the papacy in 1623, he answered to no ruler but God. Decades before Louis XIV, he had set himself up as a Roman Sun King, well aware that astronomers like Galileo were arguing that the sun was the center of the universe. Vain by nature, proud of his poetic talents and his blandly handsome face, Urban had come to imagine his pontificate as an event of cosmic importance, ordained at the beginning of time by God Himself and literally etched on the heavens in the movements of the stars.[1] Generous in his patronage, the pope cultivated an international coterie of artists, writers, and musicians who were ready to extol him in story, song, statue, and fresco, along with the members of his large, grasping family. For all his faults, moreover, the pope had daring, refined taste. The artists, architects, and musicians who worked under his sponsorship were soon creating a distinctive Roman version of baroque style, combining classical elegance with sensual appeal. The rich surfaces of Rubens's oil paintings, Bernini's sculpture, Borromini's architecture, and Caccini's music all built their outward luxury of texture and sound on powerful basic structures of composition; Barberini Rome perfected the technique of delivering a clear message in a sumptuous package.[2] And the message conveyed by all these works of art, however sunny, was one of absolute despotism. Despite his own origins among the courtly gentlemen of the Florentine academies, Urban now stood at the head of a theocracy. He presided over a state in which the punishment of heretics was still a thriving form of public spectacle, where intellectuals labored under the shadow of the Inquisition and its Index of Prohibited Books — and where working as a censor for the Inquisitors was one easy way for a scholar to find employment in the Eternal City.

Religious orders played an inordinately important part in Roman intellectual life. Urban had already shown the Dominicans particular favor by remodeling their house at Santa Maria Sopra Minerva on a grand scale, but he also counted on the Society of Jesus, every one of whose fathers had sworn him a special vow of absolute obedience.[3] The Jesuits' loyalty, and the cast of mind it represented, had won them untold influence as they began their second century of activity. They had set a new university, the Collegio Romano, and a massive new church, Sant'Ignazio, in the heart of the city, where the ancient Roman road later known as the Via del Corso met the medieval Via Papalis. From this bastion, across the street from the Dominicans, they exerted their dominion over Roman cultural life in Curzio's day through commanding figures like the architect/scientist Orazio Grassi and the mystic polymath Athanasius Kircher, sent in to teach astronomy at the Collegio Romano in the immediate aftermath of Galileo's trial.[4] Many of these

47

learned priests engaged in lively conversations with members of Rome's academies, most notably the exclusive, scientifically oriented Accademia dei Lincei, founded by the young Roman nobleman Federico Cesi in 1603; Galileo's pride in belonging to the Lincei was so great that in his *Dialogue on the Two Great World Systems*, he referred to himself simply as the "Accademico Linceo." But there were other academies as well: the Desiosi, the Umoristi, the Oziosi of Naples, many of whose members also maintained an important presence in Rome. Jesuit researchers swapped experiments on the earth's "athmo-sphaera" with Evangelista Torricelli, who could be found busily at work inventing thermometers and barometers as he tried to create a natural vacuum. Glass tubes hung out the windows of the Collegio Romano to explore how siphons worked, and the observatory that Orazio Grassi put in place on the top story of the Collegio still acts as one of Rome's important weather stations. In matters of art, literature, history, and natural philosophy, gentlemen, lay and religious, argued back and forth about empirical evidence and biblical revelation, often to their mutual delight. But the Jesuits were also remarkable in directing their close attention to the intellectual capacities of women, and were rewarded by faithful support.

Always, however, the religious orders' ability to overpower dissenting opinion went far beyond the conventional weapons of academic debate. In Rome's learned halls, whether at the Dominican stronghold of Santa Maria Sopra Minerva, at the Jesuits' Collegio Romano, at the University of the Sapienza, or in the private academies of the Lincei, the Desiosi, or the Umoristi, the threat of heresy could emerge at any time, and hence the stakes of learned controversy were incalculably higher than at the court of the Medici or in the gentlemen's academies of Tuscany.

Tuscan academicians came from largely similar backgrounds and obeyed the same basic set of social rules. Papal Rome, on the other hand, drew from an international population of scholars, many of them in contact with India, China, and the New World. Only the Paris of Louis XIV could compare in Curzio's day with the cosmopolitan sophistication of Pope Urban's court. Unlike scholarly discussions in Tuscany, debate among these most privileged of *litterati* proceeded with less careful attention to details of behavior and more attention to the overt manipulation of power. The sophistication and allusiveness of Roman controversy were keyed up still further by the vigilant eye of the religious censors. Scholarly debate in the papal court, to an exquisite degree, could be quick, smart, and nasty.

In 1632, as polished a diplomat as Galileo, in spite of all his courtly skills, misjudged the tenderness at the heart of the Barberini pope's sense of honor. Until that year, the two men had been friends of long standing, and Galileo's view of the solar system held many points of attraction for a pontiff whose personal symbolism was so obsessed with the themes of sun and light. Those themes had found expression in a strikingly novel fresco, Andrea Sacchi's *Allegory of Divine Wisdom*, commissioned to adorn an important reception room in the Palazzo Barberini, the papal family's new dynastic stronghold on the Quirinal Hill. There Sacchi, adhering to an intricately detailed written program, presented the pope's many virtues as constellations perched in the heavens. Behind them an immense yellow sun beams with blinding intensity while a big blue-gray Earth hangs precariously just above viewers' heads. In the universe of Sacchi's painting, this low-down Earth is nowhere near the center of attention, and it is hard to believe that the patrons who commissioned such a fresco put the Earth at the center of any universe, painted or real. Cardinal Francesco Barberini, the most intellectual member of the pope's family, almost certainly agreed with Galileo about the structure of the cosmos—the solar system—and must have done so with the tacit blessing of his uncle, the pope.[5]

Then Galileo, a man with a certain vanity of his own, published his *Dialogue on the Two Great World Systems*, printed in Florence in 1632. There, in the mouth of the dull-witted pedant Simplicio, he paraphrased one of the arguments by which the pope, in a private conversation with Galileo, had once attempted to reconcile the Galilean solar system with traditional biblical cosmology. Urban suggested on that occasion that God could have designed the universe any way He wanted, whether the design conformed to human—that is, Galileo's—ideas of consistency or not. Galileo's effort to find a better explanation than traditional cosmology for such natural phenomena as the tides was, therefore, misguided from the outset. In Galileo's *Dialogue*, Simplicio put the same argument this way to his friends Sagredo and Salviati:

I know that both of you, if asked whether God in His infinite power and wisdom could have endowed the element of water with the reciprocal motion we observe in it in some other way than by moving its container, I tell you that I know what you will reply, namely that He would have been able and knew how to do so in many ways, including ways that were inconceivable to our own intellects. Hence I conclude straightaway that if this be so, then it would be utter audacity for someone else to limit and confine divine power and wisdom to some personal whim.[6]

The pope's contention—that by insisting on a logical but unorthodox explanation for what he observed in nature, Galileo was subjecting divine omnipotence to human limits—was feeble, but its intentions were earnest; as a head of state,

Urban had to contend with earthly matters like the Thirty Years' War as well as the structure of the heavens. He may have hoped, at a time when Christians were regularly killing each other over points of doctrine, that Galileo could help to resolve questions about cosmology without resorting to schism and violence. For Galileo, however, the pope had entirely missed the point: drawing a conclusion from evidence had nothing to do with putting human limits on God's power; rather, it was using God-given common sense to gain better understanding of the God-given universe. Besides, the great astronomer had his own complement of Florentine arrogance.[7]

The *Dialogue on the Two Chief World Systems* puts Simplicio's restatement of the pope's modest proposal on the very last page of four hundred, all of them packed with sharp repartee and brilliant reasoning, as if Galileo assumed that Urban would never read the book from beginning to end. To make matters worse, after Simplicio has ventured his point, Salviati, the character modeled on Galileo himself, taunts the poor pedant with a sarcastic comment: "A wondrous and truly angelic doctrine!"[8] Not surprisingly, the pope, who had in fact read Galileo's book from beginning to end, took mortal offense. He also took revenge, as only a pope could do: he called in the Inquisition. The Inquisitors soon discovered that Galileo had already been issued a warning in 1616 to cease and desist from teaching or talking about the solar system, and so the old astronomer, at sixty-nine and in failing health, went to trial.

To the Inquisitors themselves, the health of the church depended on reverence for Christ's vicar on Earth. Galileo, for all his intellectual glory, never stood a chance of escaping punishment. To bolster his ideas, he had repeatedly offered his own interpretations of the Bible, defying a proclamation of the Fourth Session of the Council of Trent: "Furthermore, in order to control unruly minds [the Council] decrees that no one, trusting in his prudence, should, in matters of faith and morals that pertain to the edification of Christian doctrine, dare, twisting the holy scriptures to his own ends, to interpret the meaning of these same holy scriptures in a sense counter to that which holy mother church has upheld and upholds."[9] From beginning to end, however, Galileo's trial also bore the stamp of Urban's personal pique. The pope attended all its sessions and watched implacably as the gouty old man was forced to recant on his knees:

> I Galileo, son of the late Vincenzo Galilei of Florence, aged seventy, facing judgment in person and kneeling before you Most Eminent and Reverend Cardinals, general Inquisitors against heretical depravity in all the Christian Republic, having the Holy Gospels before my eyes and touching them with my own hands, so swear that I have always believed, now believe, and with God's help shall in the future believe everything that the Holy Catholic and Apostolic Church preaches and teaches.
>
> But . . . after having been warned juridically by this same Office that I should by all means abandon the false opinion that the Sun is the center of the universe and that it does not move, and that the Earth is not the center of the universe and that it does move, and that I could not hold, defend, or teach said false doctrine in any form, neither in speech nor in writing, and having been notified that said doctrine is contrary to the Holy Scriptures, I have been judged vehemently suspected of heresy for having written and handed over for publication a book in which I treat this same already-condemned opinion and adduce arguments in its favor . . . with sincere heart and unfeigned faith I abjure, curse and detest said errors and heresies.[10]

This recantation saved Galileo's life; he was convicted of heresy but sentenced to perpetual prison instead of death at the stake. Of the ten cardinals who oversaw the trial, three refused to sign the final sentence; one of them was Francesco Barberini.[11]

Most of the people who now sat down to read Curzio Inghirami's *Ethruscarum Antiquitatum Fragmenta* had followed
Galileo's trial closely; several were directly involved. Cardinal
Francesco Barberini received a copy of the book for his library,
where it caught the attention of his friend and secretary, Cavaliere Cassiano dal Pozzo. Another copy went to the powerful
Cardinal Giulio Sacchetti, who read the text and passed it for
comment to a fifty-two-year-old legal scholar in his entourage,
Vincente Nogueira, whose long experience in the archives of
Spain and Portugal gave him particular expertise in working
with old documents. This expert analyst, who Italianized his
name as Vincenzo Noghera, also had another job: spying for
Portugal, which had been annexed by the Spanish crown in
1580 and would shortly regain its independence (in 1640).[12] In
Malta the scarith caught the attention of the thirty-six-year-old
papal legate, a Sienese cleric (and future pope) named Fabio
Chigi, who had just come across a cache of Etruscan inscriptions himself.[13] Carefully, Chigi noted down some important
information disclosed by the scarith: the foundation date for
his native Siena and the date, according to Prospero the augur's
Etruscan calculus, for the birth of Christ. Another copy went
to the Vatican Library, where readers included the Greek *scriptor* Leone Allacci and his friend Melchior Inchofer, a Hungarian Jesuit who had also been Galileo's most damaging examiner
for the Holy Office.[14]

The Etruscans were already an object of intense interest for
this circle, and the discoveries at Scornello seemed to be the
most exciting find for years, or perhaps, given the theological
and astronomical revelations in Prospero's texts, of all time.
Among them, it was gout-stricken Vincenzo Noghera, sick in
bed at Cardinal Sacchetti's palazzo in Bologna, who first dared
to speak his mind. A legal scholar by profession, Noghera was

too scrupulous to air any of his conclusions on personal authority alone. Instead, in a letter to Cardinal Sacchetti, he couched his thoughts in the form of a syllogism:

> If I prove that these Antiquities were not written at the time they say they were, nor that it was possible that they could have been written some one hundred twenty years later, it follows that they were not written or collected by the Author [Prospero] whom they name, and consequently, it is all a fraud, and a fiction, that has nothing good about it.[15]

Noghera also confessed his doubts to Cavaliere Cassiano dal Pozzo, a minor aristocrat of vast erudition who acted in Rome as the agent, secretary, and confidant for Cardinal Francesco Barberini.[16] Neither Noghera nor dal Pozzo needed to see a real scarith in order to raise the usual doubts about Etruscan paper; they both agreed that so far as they knew, paper was a fairly modern invention, and that a real Etruscan augur would have written his books on cloth.[17] Furthermore, they found a number of inconsistencies in the scarith texts, enough to make each begin to think that the prophecies of Christ's birth, the sad story of Scornello's conquered citadel, and the figure of Prospero the Etruscan were all too good to be true. Cassiano shared Noghera's firm suspicion that the scarith had to be a forgery.[18]

By the time Cassiano and Noghera began discussing the Volterran Antiquities in 1637, Curzio Inghirami, at twenty-two, had reached the age of legal majority. To Cassiano dal Pozzo, twenty-two still seemed far too early in life for the young Volterran to have mounted so elaborate a hoax himself. The scarith had been found deep in the ground, entangled in webs of tree roots; if Curzio had discovered them at nineteen, they must have been put in place when he was only a boy. Furthermore, Cassiano saw the sophistication of the buried cocoons and their trilingual texts as the work of an older, more polished, and more perverted intellect, and proposed a possible culprit: Guillaume Postel (1510–1581), a French scholar of Semitic languages who had spent his early career in Tuscany. In 1551 Postel had published a dubious essay with the ponderous title of *Commentaries on the Origins, Institutions, Religion and Customs of the Region of Etruria, which was the First to be inhabited in the European World, and especially on the Doctrine of the Golden Age*.[19] Dedicated to Grand Duke Cosimo I, the book had clearly struck the monarch's fancy with its assertions that the Etruscans had practiced the worship of the true God until their *prisca theologia* was undermined by Greek and Roman paganism. Postel's booklet was perhaps more fiction than an outright forgery, but it was still notoriously unreliable. Some of the Frenchman's later work made no less liberal use of spurious sources and forced conclusions, while Postel himself—endlessly restless, prone to apocalyptic visions and strange theories, prosecuted on occasion for sorcery—had certainly been odd enough, and unscrupulous enough, to have planted a scarith or two during his years in Tuscany.

Vincenzo Noghera, however, would have none of Cassiano's theory. As he admitted in his letter to Cardinal Sacchetti:

I'll never believe that Postel was the Author, not because one could not suspect every sort of wretchedness and chicanery on his conscience, given that he was tried for Magic and Sorcery—and I don't know whether or not he was convicted—and was also a man of bad life and bad habits, but because I have great Familiarity with the depth of his knowledge and learning, for I have read many of his books, including the prohibited ones . . . in all of which Postel has demonstrated very sound learning, while in the Oriental languages he is a veritable prodigy. . . . Postel would lie like a person of consequence, not like a fool—and there is not a page to be found here in which mistakes cannot be detected.[20]

Indeed, Noghera had his own candidate for the inventor of the scarith. He never named that person outright; a clever lawyer had no need to. His language was explicit all the same:

Although the number of Historians and other Interlocutors introduced here is large . . . there is not the slightest difference in style to be discerned among them, but rather such uniformity and similarity of sound that they might all have lived at the same time—or, to say it better, they are all written by a single hand, by the same pen, and it is all juvenile work.[21]

Just to make certain that Sacchetti would understand him, he drove the point home a second time. First he quoted the scarith text that appeared on page 3 of the *Ethruscarum Antiquitatum Fragmenta*, in which Prospero described how he had assembled his cache. It was a particularly haunting document because in many places the ancient paper had simply crumbled away, leaving the narrative with tantalizing holes:

With great labor I gathered the prophecies about the citadel, about Volterra, about me, about these scriptures: I knew that the citadel would be wasted, and that I would either be taken captive by the Romans or be killed, but because I would bring such joy to him who would be born in Tuscia after many centuries, whose father would own the remains of this citadel, he would have a faithful friend; he, his friend, and his father would be students of antiquity, but . . . he would read . . . himself . . . must be written, and re . . . he will find these scriptures; if by chance another shall find them and touch them, let him not dare; otherwise he will die miserably, victim of the wrath of the Gods Above and Below.[22]

Noghera, however, managed to read this testimony utterly unmoved. To Sacchetti, he scoffed:

Just see whether the Gentleman who found this Treasure could have described himself more patently if he had referred to himself by name, what with all these distinguishing signs: having a Father who is Lord of the Castle, and a man of quality, being one himself, along with his friend.[23]

In legal terms, Noghera went on to observe, "speech providing a detailed description may be taken as equivalent to naming the person under discussion."[24] It would have taken a more feeble intellect than Sacchetti's to have missed Noghera's point: to his mind, the forger of Scornello was none other than Curzio himself.

With unusual indiscretion, Cardinal Sacchetti sent a copy of Noghera's letter straight to Scornello. As one of the most powerful men in Europe, the cardinal may not have considered clearly enough what repercussions his actions might have on a more vulnerable soul, as Noghera certainly was in the late 1630s: an exile and a spy who had lived for some years in Italy under an assumed name, Francisco de Noya, that disguised his family's Jewish origins. Indeed, it is clear from the frankness of his missive that Noghera never intended for it to circulate widely. A copy certainly came into the possession of Fabio Chigi, Cardinal Sacchetti's close friend and protégé; another copy, perhaps two, went to Cardinal Barberini and Cassiano dal Pozzo; and one, of course, had gone to the Inghirami, but otherwise the letter seems to have been more talked about than read.

Quickly, Curzio drafted Noghera a tart reply, addressing some of the objections about paper and chronology, and once again professing his own good faith as the scarith's discoverer and exegete. Cavaliere Giulio took a different approach; from that moment onward, he mounted a ruthless, and ultimately successful, campaign to bully Noghera into silence.

From his own vantage, Cavaliere Giulio may have felt that Noghera had betrayed him; the two had been friends, and Noghera would continue to describe himself as a "client" of the Inghirami family. Writing from his sickbed in Bologna, the learned spy may not have realized how deeply Cavaliere Giulio was embroiled in the affairs of Curzio, who was, after all, only a distant nephew. But involved Cavaliere Giulio was, and ready to defend the Volterran Antiquities from every adversary. With ruthless precision, he assailed Noghera's reputation at what seemed to be its most vulnerable point, and because Noghera was involved at the same moment in genuinely subversive activities—Portugal would break away from Spain two years later, in 1640—and already suffered cruelly from his chronic illness, the turmoil of the next year's wranglings with the Inghirami would send him into an extended agony of body and spirit.

Vincenzo Noghera held a hereditary knighthood from the Grand Duchy, first granted his grandfather and his father by the Grand Duke Cosimo I, and awarded to him personally by the subsequent grand duke, Cosimo II (the present Grand Duke Ferdinando's father).[25] The social status conferred by this degree was as important to Noghera and his family as the analogous Knighthoods of Saint Stephen were to the Inghirami, and it was on this point of honor that Cavaliere Giulio mounted his attack. He threatened to bring Noghera to trial in Florence for injuring the majesty of Cosimo I. In a certain sense, the idea was absurd; Cosimo had been dead for decades and was presumably beyond lèse-majesté, insult, or injury. But Cavaliere Giulio may have known that Noghera had a few skeletons in his closet as well: he had escaped to Italy from Spain after running afoul of King Philip IV's formidable right-hand man, the Count-Duke of Olivares, whose accomplices had still managed to cast Noghera into prison on his arrival.[26] Noghera never specifies the nature of his offense against the Conde-Duca, an offense that also alienated his own brother. Neither is it clear whether his efforts for Portuguese independence resulted from his treatment by Olivares or caused it. He was not, however, a loyal subject of the Spanish crown. Furthermore, his talents quickly brought him into contact with the most powerful cardinals in Rome. A note in a Vatican manuscript describes him as "most erudite, but he took too much pride in his knowledge and was too free in his habits."[27] Those free habits included imploring Cardinal Francesco Barberini to put back the big armillary spheres that had decorated the family library until 1633 with their models of the Copernican solar system—"just put up a label that says 'the damnable Copernican system,'" Noghera advised.[28]

If Giulio Inghirami's lawsuit was meant to intimidate this clever, elusive character, it worked; the poor man was driven to despair. He was doubly shocked, as he wrote Cassiano dal Pozzo in the summer of 1638, because such behavior seemed out of line among gentlemen:

I am a client of the Lords Inghirami and a great friend of Cavaliere Giulio and all of their train, but I was very displeased that they wanted to bring in, without need and without propriety, the memory of that Hero [Cosimo I]. If they were not gentlemen and noble persons, I would have thought that they had done so in bad Faith, but because they are who they are, I prefer to think that they truly believed his Highness to have been offended.[29]

Noghera's correspondence shows that by 1638 the scarith had become more than an archaeological problem. Debate about their authenticity had become a political fight between Tuscany and Rome. Behind Vincenzo Noghera and Cavaliere Giulio loomed the figures of Cardinal Sacchetti, Cardinal Barberini, and the Grand Duke of Tuscany, as Cassiano dal Pozzo learned from Noghera in the same letter, and probably knew perfectly well himself:

I have received various letters from Rome advising me to explain my position to the Grand Duke, and certainly I would have done so if I were in better health, and not so much because I think that in that Olympus such patent errors would make any impression, as to re-instate myself in the favor that my family has so worthily obtained, and to reverse the import of the accusation with which he intended to charge me. I kiss Your Lordship's hands for the favors you have done me with the Lord Ambassador Count Nicolini and the Lord Bailiff [Andrea] Cioli, the Secretary of State, and I beg Your Lordship that if it seem convenient to you (necessary I know it is not) [please write] a letter of the most humble obligation to His Highness [the grand duke] . . . so that I will not damage the degree I have from his Father, and my father from his Grandfather. . . .[30]

Eventually, some hurried negotiations between Florentine diplomats and Roman grandees stopped the threatened lawsuit against Noghera; its substance was questionable in any case. Cavaliere Giulio resorted to reminding Noghera of their old friendship, and then, at last, offered a monetary settlement. In August 1638 Noghera again wrote Cassiano in deep exasper-ation, though perhaps he protested a bit too much:

I always meant the pungent expressions impersonally, and *in rem*, with-out imagining even a venial sin in Signor Curzio (excessive credulity, yes); otherwise they would have stayed in my pen, because I have the greatest respect for that most noble family.[31]

The heat of summer, the pressures exerted by Cavaliere Giulio, and his own miserable health finally exhausted Noghera's pa-tience. He agreed (for a price) never again to mention the scarith. As he explained to Cassiano in September 1638:

I have repented a thousand times ever having taken up a pen about the antiquities of Volterra. It should have been enough for me to satisfy my obedience to the Lord Cardinal by telling him orally my sense of it, although he liked my essay as such, and when he asked my permission to show it to the Signori Inghirami, I happily loaned it to them, for I never suspected a fraud of those Signori, and so I never thought that they would try to exonerate themselves of such an accusation, but as soon as I realized the contrary, I obtained from the Lord Cardinal not only the original letter but also the reply and I have kept both under lock and key without making any copies.[32]

And, indeed, Noghera seems to have been as good as his word. None of his later letters to Cassiano ever mention the scarith of Scornello, though his adversary Cavaliere Giulio died not long thereafter, on 7 May 1639 in Florence.[33]

---- ⊰[IV]⊱ ----

About Paper

1635–1640

Tene risum, si potes.
Control your laughter, if you can.
LEONE ALLACCI[1]

As Curzio Inghirami admitted himself, his presentation of the scarith at the University of Pisa had inspired lively debate. The most outspoken skeptic on that occasion had been the university's professor of humane letters, Paganino Gaudenzio.[2] As an immigrant from the Swiss canton of Ticino, Gaudenzio exemplified Pisa's long-standing tradition of offering chairs to gifted foreigners. His formidable skills in Greek, Latin, and Tuscan vernacular had brought him down to Rome as a child prodigy and to his post at Pisa in 1628. Arrogant, vain, irascible, and ravenous for fame, Gaudenzio was also a scrupulous scholar. By August 1636, well before Curzio's *Ethruscarum Antiquitatum Fragmenta* had left the presses, the professor from Pisa had himself prepared a manuscript for printing in Florence. By Sep-

62

tember, *De Charta: Exercitatio* had obtained approval by both the Florentine Inquisitor and the civil censor.[3] The essay combines copious references to ancient authors with simple outrage that anyone could take the scarith seriously. Its preface sets the tone:

It is quite remarkable that people can be found today who dare to affirm that among the Romans of Sulla's time, paper such as that which we have was not unknown, made from torn and macerated pieces of linen fabric.[4]

With careful attention to detail, *De Charta* then traces the use of the word *charta* through antiquity to conclude that among the ancient Greeks and Romans, it had always signified writing surfaces prepared from Egyptian papyrus. "I'll cry out, and again I'll cry that the *charta* of the ancients was made of papyrus," Gaudenzio exclaimed in conclusion.[5] Only when papyrus fell out of use in the Middle Ages was a substitute invented, and Gaudenzio explained that "most appropriately, then we called our paper *charta*, because it has great similarities to the *charta* that had been made from papyrus."[6]

At the same time, Gaudenzio began to prepare a longer and more specific analysis of the scarith themselves (figs. 16 and 17). He addressed the manuscript of *Paganino Gaudenzio's Animadversion against certain Antiquities published under the name of Prospero of Fiesole* (*Paganini Gaudentii in Antiquitates quasdam editas sub nomine Prosperi Faesulani Animadversio*) to a Tuscan bishop, Francesco Venturini, reviling the Etruscan Antiquities with undisguised glee, from an early "Who will not admit to Prospero's fiction? Who will not explode his audacity?" to a closing "Oy, reader, oy, enough already."[7]

Gaudenzio also proposed a new candidate for the forger: Tommaso Inghirami, a scholar, actor, and orator of scintillating talent who had served as Vatican librarian at the height of the

7

ETHRVSCARVM
ANTIQVITATVM
FRAGMENTA.

LIBER PRIMVS.

Vulterrana hęc monumenta (veri theſauri Ethruſcorum) in quibus ſunt Vulter-
ræ, & aliarum Ciuitatum primordia, augmentum, potentia, & Regnum,
in hoc loco repoſuit ille Proſperus, qui alia repoſuit.
Si incidantur, mala portendunt.

AGNVS Pa
ter Vandi-
mon, qui à
Latinis Ia-
nus, à Siris
Noà vocatur, poſt aquas
relictis in Armenia, & Si-
ria Semete eius primoge-
nito, & Came ſe tertio filio
apud Nilum, aduenit in
hanc regionem cum ſecū-
do filio Iapèto, & illius fi-
lijs, & cum peruenisset ſu-
pra hunc montem, ſibi cō-
modum, poſteris iucun-
dum putauit; quare in ſu-
periori parte, quæ ſalu-
brior eſſet, Ciuitatem
ædificauit, & Cethim ap-
pellauit. Verum an. ccxx
à Cethim ædificatione Ce
thim filius Iauanis, & pro-
nepos magni Vandimo-
nis assumptis, ſecum dua-
bus Colonijs mare ingreſ-
ſus, cum diù nauigaſſet,

peruenit tandem ad Inſulam, quam ſuo, & Patriæ nomine Cethim appella-
uit. Hæc hodie Ciprus dicitur. Deinde Græci Vrbem noſtram, quæ Ce-

B ij   thim

LI.
Scarith
Die xix.
Septembr.
MDCXXXV

A

Cum ha-
ud commo
dè hi per-
legerētur
Charaćte-
res, repe-
tendos du
ximus.

B

C

16 : Latin text from Scarith #51, copper-plate engraved image, woodcut frame, and
movable type. Curzio Inghirami, *Ethruscarum Antiquitatum Fragmenta* (Florence: Massi,
1636), 7. Photo: Antonio Ortolan.

Renaissance in Rome. Born in Volterra
in 1470, this Inghirami forebear had
spent his childhood in the Florentine
court of Lorenzo de' Medici. He then
moved to Rome in his teens to serve
Cardinal Raffaele Riario and to take
advantage of the university's new cur-
riculum in "humane studies" (*studia hu-
manitatis*), which really meant an extraordinary grounding in classical languages, literature, and—uniquely for
the late fifteenth century—archaeology.[8] In 1486 Inghirami and a troupe of
his fellow students, subsidized by Cardinal Riario, staged Rome's first modern performance of an ancient tragedy,
Seneca's *Phaedra*, with Tommaso cast
as the lovesick queen.[9] To create the
stage set, he and his fellow students
followed the instructions of the ancient Roman writer Vitruvius, but they

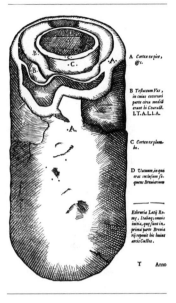

17 : Cutaway view of Scarith #98. Curzio
Inghirami, *Ethruscarum Antiquitatum Frag-
menta* (Florence: Massi, 1636), 141. Photo:
Antonio Ortolan.

proved better scholars than carpenters: during the performance, a piece of scenery collapsed in the middle of one of Phaedra's soliloquies. Resourceful Tommaso carried on with queenly
aplomb, making up new lines of Latin verse until his friends
could pull the set back into place.[10] A star was born that afternoon, and so was a nickname: Tommaso Inghirami was "Fedra"
ever after, and still is—for more than five centuries now, generation upon generation of Inghirami boys have carried on the
name "Tommaso Fedra."[11]

More than a century after his death, Tommaso Fedra Inghirami's reputation for extravagant behavior was still vivid: with
his wall-eyes, his resonant voice, the ponderous body he moved

18 : Raphael, *Portrait of Tommaso Fedra Inghirami*, circa 1510. Florence, Galleria Palatina, Pitti Palace. Photo: Alinari/Art Resource, NY.

with such surprising grace, and the startling inventiveness of his speeches, he had made friends with cardinals, popes, courtesans, scholars, handsome young men, and the painter Raphael, who painted him in about 1510 when they were both working closely with Pope Julius II (fig. 18). Paganino Gaudenzio suggested that Tommaso Fedra had planted his great cache of scarith in a grove of young oaks at the Inghirami villa, carefully putting one of them under a thinner cover of dirt so that it would be discovered first: the very scarith, wrote Gaudenzio, that an unwitting Curzio would pull from the riverbank some hundred and twenty years later. Or else Curzio was failing to tell the whole truth:

It is highly credible that some kind of memorandum was passed on by Tommaso Fedra, who seems to have been architect of these writings, to the Inghirami, the one that tells where the first paper should be sought, so that it in turn would lead the excavation of the others.[12]

In addition, however, Gaudenzio suggested that Curzio (to whom he only referred obliquely as "the Defender") might have "improved" upon parts of Fedra's hoax, and done so (such was Gaudenzio's ego) to meet the objections that Gaudenzio himself had raised when the scarith were first presented in Pisa:

One might perhaps suspect that a bit had been added to this scarith by the Defender himself, after I had said that our paper was unknown to the Romans.[13]

No matter their origin, Gaudenzio declared, the Etruscan Antiquities were not Etruscan, and accepting them as genuine defied common sense:

And if the Defender of this affair should discover a person who places trust in him, or merely fails to suspect him of trickery, then I will immediately rush to embrace this great collection of the most improbable tales.[14]

Paganino Gaudenzio probably hoped to publish his essay to great fanfare in Tuscany. But before that could happen, someone within the Grand Duchy let him know that a loyal Tuscan left the scarith and their Defender alone; they enjoyed protection from the highest ranks of the Medici court. Gaudenzio was not, on the whole, the sort to surrender to pressure; his big, assertive handwriting, devouring page after page in an exuberantly illegible scrawl, is not the work of a submissive character. He was, however, a foreign professor at Pisa, an attractive post that he had every intention of keeping.

Gaudenzio thus became caught between conflicting loyalties. The Tuscan universities had been conceived for a more specific purpose than the general enhancement of learning: in them, the Grand Duchy's intellectual superiority was put on display, whether it might be Galileo championing his new science or literary professors championing the supple beauty of the Tuscan tongue. Foreign professors like Gaudenzio served to demonstrate the universities' ability to attract the best scholars in the world. As a professor of literature, Gaudenzio was also expected to lend his authority to promoting what the Grand Duchy called "Etruscan letters."

Indeed, the residents of granducal Tuscany prided themselves upon commanding a living relic of the ancient Etruscan language. Tuscan humanists had called their vernacular idiom "Etruscan" since the fifteenth century, in the sincere belief that it had evolved from remnants of the Etruscan language mixed in with colloquial Latin. In the mid-sixteenth century, the very first grand duke, Cosimo I, had used this putative connection between Tuscan vernacular and Etruscan language to help pull his new state into a coherent cultural unit, reminding the newly subjugated communes of the Grand Duchy that ancient Etruria had been made *e pluribus una*, twelve cities united in one Etruscan League.[15] To band together was not to forfeit independence; at least this is what Cosimo wanted his more restless dominions to believe. Their common dialect provided Cosimo's subjects with another proof of their essential unity, especially when court intellectuals worked hard at driving home the contrast between the fresh immediacy of "Etruscan" language and the ponderous court Latin of papal Rome. In Cosimo's wake, speaking vernacular had become of point of honor among Tuscan intellectuals. The Accademia della Crusca urged the use of words with Etruscan roots. Most of these were sheer invention, but some, like *lettera*, *persona*, and *histrionico*, were genuine Etruscan legacies (and suggest a good deal about the quality of Etruscan life).[16] The academy's activities aimed to show that the "Etruscan" language could equal the expressive range of Latin on every subject for every occasion, from Galileo's observations of the cosmos to the government's diplomatic correspondence with the world's rulers.

When Paganino Gaudenzio began to agitate against the scarith, the Grand Duchy, still smarting from the trial of Galileo, wanted no more blows to its cultural prestige. The state had fallen on hard times since the glory days of Cosimo I, and an outbreak of bubonic plague in 1630–31 pushed years of economic decline into free fall.[17] The Medici grand dukes may once have supplied the monarchs of France with their queens, but when Grand Duke Ferdinando went looking for a wife, his chubby, pious bride, Vittoria Della Rovere, commanded only the little Italian city-state of Urbino.[18] Paganino Gaudenzio had been one of the court dependents who were called upon to make this provincial alliance look as illustrious as the famous Medici weddings of 1589 and 1604, as if poetry and art alone could compensate for the grand duke's depleted treasury and the wedded couple's staggering physical unattractiveness.[19] Cultural prestige, or its vestiges, had become one of Tuscany's only abundant commodities, and in the troubled 1630s, it was not to be trifled with.

As Gaudenzio discovered, more slowly, perhaps, than he should have, the scarith had come to stand for Tuscan antiquity in a way that transcended any concern about their quality as an antiquarian find, or indeed their authenticity. In the first place, continuing controversy kept them in the public eye in ways that a single discovery could not—not even a wondrous bronze like the Chimera discovered at Arezzo in 1549 or a lordly tomb like the tumulus unearthed outside Castellina in Chianti in 1507. Discreetly but reliably, the grand duke had adopted Curzio Inghirami as a cause; when Paganino Gaudenzio called the youth "Defender of the Etruscan Antiquities," he used a title that had almost become a professional designation. While Curzio's detractors outside Tuscany grew in number and importance, his allies within the state became more and more distinguished.

However, Gaudenzio had no intention of keeping silent. He simply sent his manuscript abroad, to a Danish scholar, Heinrich Ernst, who had seen the scarith himself during a trip to Rome. Together they decided to publish a joint work, anonymously, with the Protestant printer Johann Jansson in Amsterdam; the biographical details that appear in this *Against the Etruscan Antiquities* (*Ad antiquitates etruscas*) belong to Ernst, safely ensconced in the Protestant north, but the text itself was a joint creation.[20] Ernst used Gaudenzio's *Animadversion* virtually in its entirety, preserving its structure of preface, thirty short chapters, and afterword; the afterword was Gaudenzio's, verbatim, but the preface showed that Ernst could sling invective as gleefully as his colleague in Pisa:

It is hardly credible with what joy I opened this book, inspected it, and read it, or with what outrage I closed it, dismissed it, and cursed it. For I grasped immediately that these fragments were not written in the time of Sulla, as they wish to appear, but are counterfeit and created to hoodwink mere mortals. The matter speaks for itself.[21]

On occasion, Ernst bolstered Gaudenzio's arguments with citations from *Ethruscarum Antiquitatum Fragmenta* and from ancient authors. He added an additional chapter to Gaudenzio's thirty, taking the opportunity to praise his countryman Tycho Brahe while he ridiculed the astronomy of the scarith:

Who, unless they have their brain in their feet, does not see that all the things he has collected are ridiculous, made up, twisted? On page 4 of the *Fragments* he refers to three stars, Caris, Mor, and Turg: What, I ask, are these stars? . . .

People now, thanks to our great Tycho Brahe, draw and describe this constellation. . . .[22]

From his remote northern vantage, he could also take much more explicit aim than Gaudenzio at Curzio himself:

Nonetheless, I believe that this most learned man, Curzio Inghirami, thinks that all these things he has published are as true as if he had made them up himself, but he diverges greatly from my own opinion. For it is all too well known that many other things of this sort have been invented by citizens of Volterra. . . .[23] Why Inghirami, warned repeatedly about this matter, should want to prostitute his erudition, I cannot wonder enough. It was not, at least in my opinion, the prudence of a good man to entrust his candid simplicity to these false and fallacious papers.[24]

Curzio's verbal defenses against his adversaries seem to have been more successful than those he put to paper. Vincenzo Noghera had noted, with a whisper of disapproval, that the young man's replies to his letter had taken too defensive a tone for fruitful scholarly discussion:

> I enjoyed Signor Curzio's reply, because a certain sourness of his neither suspended nor blinded my judgment; on the contrary, I saw much learning and great erudition, and the objection about the paper on which we write today is very well satisfied. I thought that I might write him about some points that still remain weak, but [I was too ill].[25]

Fortunately, Inghirami found a kind tutor in the lively person of Benedetto Buonmattei (1581–1648), the illustrious Florentine grammarian and mathematician, whose manifold careers had included stints as a merchant, a clerk in the Tuscan bureaucracy, a parish priest, and a school headmaster before he was appointed professor of "Etruscan" at Pisa in 1632, and then elevated to the same position in Florence in 1637.[26] By the time he began to correspond with Curzio Inghirami, Benedetto Buonmattei had become the supreme arbiter of Tuscan literary style in all of the Grand Duchy. Elected to the Accademia della Crusca in 1627, he would be instrumental in resurrecting it again in 1640. His memberships in other academies, meanwhile, were legion: he had helped found the Apatisti, and also belonged to the Svogliati, the Infiammati, the Instancabili, the Spensierati, the Umoristi, the Pazzi, and a whole set of religious confraternities. The secret of his popularity seems to lie in the fact that he made friends with everyone, including, in 1638, a young English visitor to Florence named John Milton. He even delivered a long eulogy in the Accademia della Crusca for his cat, Romeo, whom Buonmattei praised on that occasion as "master of all the liberal arts."[27]

It may have been at the prodding of Cavaliere Giulio that Buonmattei first wrote to Curzio Inghirami about ways to respond gracefully to critics, but his advice is also intimately personal, a seasoned professor's sensitive counsel to a promising but touchy student. Buonmattei wrote in an extravagantly precious idiom—hence his success as an arbiter of style—but his basic advice was simple and sound: Behave like a gentleman, he suggested, and you will be treated like one.

Dear Sir. I would be of the opinion that when the objections do not regard anything specific, either one should not respond, or should deal exclusively with general remarks, and reserve the arguments that have been adduced by yourself to a better occasion, because you are not the author of those texts, and once you have published them just as you found them, it hardly seems your responsibility to answer every objection that might arise from every sort of personal whim. But you are already master of the situation, and whoever wishes to strip you of that mastery must not do so on the basis of presuppositions; let him rather come forward with clear and compelling arguments, and then you may respond at your leisure.

But if, nonetheless, you should wish to say something as a friend to those who, as friends, have sent you objections, I would advise you to be brief, and where the opponent does not prove his arguments, I would make no other answer than "I do not see it that way," "I think it is quite the opposite," or something of the sort. And where the arguments work to your advantage, I would concede them all, and not add a word to them.[28]

By example, Buonmattei's letter also provided Curzio with another lesson if he had the eyes to see it, and his subsequent conduct suggests that he did: in their time and place, style was essential to communicating substance.

Indeed, style would propel the most devastating attack to assail the *Ethruscarum Antiquitatum Fragmenta*. It came from the Vatican Library in 1638, just as Vincenzo Noghera weighed taking his pledge of silence. Noghera told Cassiano dal Pozzo earlier in the same year that the scrappy Greek librarian at the Vatican, Leone Allacci, had pressed him for a copy of his letter to Cardinal Sacchetti; Allacci claimed to be drafting his own attack on the Volterran Antiquities and wanted to know Noghera's thoughts. Noghera apparently refused; at least this is what he told Cassiano dal Pozzo:

When I was importuned by Signor Leone Allacci, and by others to send [Curzio's] reply to them, I said that I was quite satisfied by his learned reply, to the point that I saw hardly any problem left to be resolved, and Your Lordship may take this as infallible truth, as may Lord Bailiff Cioli, and whatever the Signori Inghirami publish I do not want this letter published ever, not the least word of it, because I want to preserve the favor and friendship of those Signori and because this is what is appropriate to my status, age, fortune, and peace of mind.[29]

Noghera was rare in his discretion. The people who took up the scarith of Scornello after him can be seen at work in a contemporary portrait of the papal court called *Apes Urbanae*—"Urban's Bees"[30]—a title that plays on the three bees in the Barberini coat of arms, on the pope's own name, and on that name's conscious identification with the Urbs par excellence, Rome. The title of the *Apes Urbanae* also calls up the idea of a busy hive, an image that Urban VIII strove to cultivate for his apostolic court.[31] It was a hive into which no young aristocrat from the Tuscan provinces stepped lightly, for the industrious activity of Barberini Rome had its stinging wit as well, and nowhere, perhaps, more than in the person of the author of the *Apes Urbanae*, Leone Allacci, born, like Homer, on the island of Chios but long transplanted to Italy. In 1637, as a scriptor at the Vatican Library with command of several languages as well as university degrees in philosophy, theology, and medicine, Allacci, at fifty-one, had just published his own set of controversial ancient texts, the purported writings of Socrates, Antisthenes, and other Socratic philosophers—essays that flew in the face of Plato's contention that Socrates had never written anything (Plato, not Allacci, was right).[32] Inghirami could not have hoped for a more knowledgeable reader, or a sounder judge of literary style—if that is what he hoped for.

Cavaliere Giulio's ruthless dealings with Vincenzo Noghera probably explain why, when Allacci finally struck at the scarith, he struck so hard. Like Paganino Gaudenzio, he published his diatribe abroad, in Paris, Rome's rival for cultural leadership in Europe. *Leonis Allatii Animadversiones in Antiquitatum etruscarum fragmenta ab Inghiramo edita* (Leone Allacci's warnings against the *Fragments of Etruscan Antiquities* published by Inghirami) came out in 1640. He reprinted it two years later, this time in Rome itself.

From his very first sentence, Allacci abandoned any pretense of gentlemanly discourse; instead, he likened Curzio's beautiful volume to a pile of manure:

A new Augean stable, full of foul odors and outrage, can scarcely slow my angry charge against ridiculous fables and pure trifles. For lately some *Fragments of Etruscan Antiquities*, friendly of title, seductive of subject, rare in their novelty, and alluring in their stately elegance, could not soon enough beguile my respite from my studies. But alas, miserable condition of things human! Should I not now say, more appropriately than ever, "Big book, big trouble"? There is poison beneath that honey; there are thornbushes and brambles everywhere. . . .
Was I not a fool, impetuous, and more merciless with myself than my very Deceiver, if, warned to beware so many times at the beginning of that book—*Cave, Cave, Cave!*—I did not beware, but helped myself to what it offered?[33]

Leone Allacci was a connoisseur of books, whose joy at working in the Vatican Library spilled over into his voluminous publications and his still more voluminous correspondence. His invective shows just how attractive the printer Amadore Massi had managed to make the *Ethruscarum Antiquitatum Fragmenta*. But beautiful form did not alone make beautiful content. Allacci was appalled at what he read:

This is clearly so far from the golden age in which our Prospero pretends to have written that even the least Historian of his era—or our own—would be ashamed; that's how ineptly most of this is managed, and awkwardly: the words, the expression, the clauses, the structure, the manner of speech, the statements, and the whole style itself.[34]

Within the confines of Tuscany, a Benedetto Buonmattei could recommend adopting a tone of calm benevolence in dealing with detractors, but Allacci's savage wordplay observed a different set of rules: in a community of gifted immigrants like papal Rome (and imperial Rome before that), where the Tuscans' sense of ancestral community competed with a swarm of other regions and other cultures, people made their reputations quickly or not at all. Allacci frankly aimed at destroying first the *Ethruscarum Antiquitatum Fragmenta* and then the reputation of their Defender:

To his readers, the Deceiver is, at one and the same time, to be pitied and to be laughed at; he incites tears and guffaws. And although my spirit shrinks from the task, and would keep its distance from gathering such matters, nonetheless, in order for the Reader to inspect the truth, I must extract a few idiocies out of the infinite number that this book gushes forth. Certainly he will take pity on me for my part; while others gather flowers from the manure, I collect, albeit necessarily, manure.[35]

After long service in defense of the scarith as physical entities, Curzio Inghirami now found himself compelled to face a new

kind of assault: pure criticism of his texts. His reprieve from
discussions about paper, ink, and linen books turned into a discussion about Prospero the Etruscan's character and social status, and hence, with barely disguised directness, about himself.

The very title of Allacci's *Animadversiones* already worked a
small but significant variation on Curzio's title *Ethruscarum Antiquitatum Fragmenta*. Curzio's spelling of "Ethruscarum"
reflected an affectation first introduced in the late fifteenth
century by the antiquarian and forger Annius of Viterbo, who
had insisted that the Etruscans had once commanded a vast
kingdom that extended the whole length of Italy; the citizens
of this wider Etruscan state spelled the names of their region
and themselves with an extra "h": "Hetruria" and "Hetrusci."
Followers of Annius, as well as some unwitting detractors,
sometimes modified this spelling to "Ethruria" and "Ethrusci,"
but in every case an additional "h" expressed belief in this
greater Hetruscan kingdom, founded by Noah/Janus and therefore not only vast in its geography, but also providentially
enlightened. The Grand Duchy of Tuscany provided the outstanding example, commemorating its own territorial con-
quests by calling itself Hetruria in Latin inscriptions; Ferdinando himself was the Magnus Dux Hetruriae.

Leone Allacci, however, even in quoting Curzio's title, carefully omitted the "h" in "Etruscarum"—and hence eschewed
any lingering homage to "Hetruria" or "Ethruria," those magic
kingdoms of sixteenth-century myth. "Etruscans, not Ethrus-
cans; Etruria, not Ethruria; Tuscans, not Thuscans," he intoned,
permitting himself only the steely concentration of a philolo-
gist.[36] Only twice did he use the word "Hetruscan," and he used
it to mean fake.[37]

Allacci, like Noghera, Gaudenzio, and Ernst before him, addressed the matter of paper in the ancient world, asserting once

again, bolstered by authors ancient and modern, that the Etruscans had written on linen books. He dissected the scarith's
chronology, comparing it with the latest in contemporary
scholarship—and with the invented chronologies of Annius of
Viterbo. He quoted Prospero's breathless comment: "I have no
paper because the siege is imminent" and pointed out:

In these wretched Scarith there are not only papers with writing, but
also many without, wrappers for the written pages, which I saw myself.
He could have written perfectly well on these. So if he had no paper,
why did he bury so many pieces of paper with nothing written on
them? Who is going to believe that paper is what runs out during a
siege?[38]

Chiefly, however, Allacci's attack cut to the heart of Prospero of Fiesole's Latin style, exposing example after example of
anachronistic usages whose monstrosity, he assured his readers,
would have made Cicero blush, despite the fact that Prospero
and Cicero were supposed to have been contemporaries. A far
cry from the cultivated gentility of a Benedetto Buonmattei,
Allacci's style stabbed and parried with ruthless directness: the
Animadversiones decry Prospero's deviant Latin as "plebeian"
and ask how any scholar of merit can believe an author who
"gushes barbarism and vulgar forms all over the place."[39] Despite Allacci's protests that "I loathe to look longer at so vile,
and muckulent a text,"[40] he roots its lapses out systematically
and holds them up for horrified scrutiny.

In gathering together these excerpts, I've long since developed a callus
on my stomach from prolonged contact, but certainly not the point
where I'd not prefer to drink bilge rather than to hear [words] from
which I recoil just as I would from a snake. Are these portents of
words, or monsters? Scarith, Caris, Mor, Turg, Asgaria, Vlerda,

Dorchethes, Lartes, Saph, Roith, Ochincres, Brocon, Spugi, Barconictus, Ancironae, Schilia, Cronuethia, Schesia, Procravia, Ocalia, Dantelia, Bentia, Porachal, Balth, Rebalth, Rurerebalth, Vosgaria, Onebrae, Enebrae, Inurnes.[41]

With his own potent command of polyglot invective, Allacci argued by example that true aristocracy was revealed by
correct Latin style rather than by ancestry. It was an old hu-
manistic argument, honed to perfection more than a century
earlier in this same Rome of ambitious immigrant scholars or-
biting a court of self-made churchmen. The argument still
worked in the Rome of Urban VIII, self-made like many of his
colleagues, and like most of Rome's self-made men, he owed his
success to his own intellect. By taking Latin style as his line of
attack, moreover, Allacci blunted the Inghirami family's ability to intimidate him, for whatever Leone Allacci's family origins, his brilliance made Prospero of Fiesole, and by implica-
tion Curzio, look like a provincial hack.

For Allacci's attack on Prospero of Fiesole's Latinity also
launched a barely disguised personal assault on Curzio Inghirami himself. By repeatedly excoriating the dead Etruscan's
style as plebeian, the *Animadversiones* implicitly rejects the
foundation of noble blood and erudition on which Curzio and
his family had so far constructed their defense of the Etruscan
Antiquities. With devious mastery, the text of the *Animadversiones* accuses only the deceased Prospero, rather than the living
Curzio, of vulgarity and sham erudition, but the flying mud
stuck to Curzio all the same:

Let anyone who wishes believe that this fogmaker from Fiesole can see
further into the deepest mysteries than other Prophets, beloved of
God, specifically elected by Him to receive this gift, and to proclaim

them more plainly. . . . If this Balaam's ass from Fiesole brays so grandly by divine guidance, it is still impossible for him to proclaim the prophesied and crucified King of the Jews; if anything, he is like those other heretics whom the Saintly Holy Fathers attacked to affirm the truth of their faith against the Gentiles, nor would he be proclaiming in so intelligible a voice; instead, he would stutter and mumble about this definite event like some foreign tourist.[42]

Like Vincenzo Noghera, but at far greater length, Allacci built up his argument on the premise that Curzio Inghirami had forged the scarith of Scornello, and pounded it home with vicious finesse. Addressing Prospero, Allacci made it clear that he was really addressing someone else:

There's a puerile placement of words in this phrase. . . .
The gods blast you, greenhorn. . . .
O scribe of the sacred college [of augurs], you evident adolescent! O blundering clown![43]

Addressing his readers at the very end of his book, he leveled a final warning:

I think that these Antiquities, with their paper, pitch, and the rest of it, are much more recent than Tommaso Fedra—and I only hope that they are not very recent indeed! Still, my spirit tells me that Curzio Inghirami, who published such things, is a cautious, prudent man, a student of true, not false and outrageous, antiquity. . . .[44]

In addition to demolishing the forger's—that is, Curzio's—execution of the hoax, the *Animadversiones* also proceeds with exquisite ruthlessness to explore the matter of its motivation. In the preface to Allacci's essay, the publisher, Sebastien Cramoisy, declared that greed will induce human beings to do strange and terrible things, including deception, but what of people who deceive for other reasons?

In the last century a certain Annius of Viterbo, in order to provide himself with fame and glory among Subsequent Generations, dared to publish, in a fraud unheard-of before that time, various ravings of his depraved mind.

Behold these brand-new, not dissimilar sycophantic *Fragments of Etruscan Antiquities* . . . when Leone Allacci sniffed out the fraud, he could not restrain himself from revealing so egregious a fraud, pulling away its mask and stripping away its cosmetics. . . .[45]

The comparison with Annius of Viterbo did more than raise the specter of a notorious forger. Allacci realized that in many respects, beginning with the spelling of the title of the *Ethruscarum Antiquitatum Fragmenta*, Annius of Viterbo had provided Curzio with his chief model. The damage inflicted on Curzio Inghirami by Allacci's slim diatribe can be measured by the fact that the *Animadversiones* went through two printings in two years, one in Paris and one in Rome; it sold internationally, and sold well.[46] As an octavo volume without illustrations, it was relatively cheap to produce, unlike the complicated interplay of type and cuts used for Curzio's *Ethruscarum Antiquitatum Fragmenta*. More damning proof of its effect is the fact that ever afterward Curzio Inghirami and his defenders would only dare reply to their critics in Tuscan vernacular. For all his reputation as a virtuoso, Curzio would never again subject his Latin style to public scrutiny.

Allacci had done more than dismiss Prospero and "the Deceiver" as crude provincials. He also implied that in comparing Curzio's personal Latin style with that of Prospero of Fiesole, he had come to conclude that the two writers were one and the same person, although the *Animadversiones* carefully avoids saying so outright. Thus Allacci served notice that any further exposure of Curzio's stylistic quirks in Latin meant further evidence on which to base an outright accusation of fraud.

For all practical purposes, academic debate on the scarith of Scornello was settled in 1640 by an essay of two hundred scathing pages. The real battle, however, had only begun.

───────────────── ⟨ v ⟩ ─────────────────

The Defender Defended

1641

In 1641, on a trip to Florence, Curzio and Inghiramo Inghirami met the German scholar Lukas Holste, a close associate of Cardinal Francesco Barberini who had been engaged by Grand Duke Ferdinando to catalog the Medici library. Hoping to gain this influential figure as an ally for the scarith, Curzio extended an invitation for Holste to come to Volterra, an invitation Holste initially hesitated to take up: caught between his friends and rivals in Rome and his temporary employer the grand duke, he could not afford to take sides in the scarith debate until he found a more permanent job.

Curzio, in the meantime, wrote to Holste to encourage a visit some other time, and to report an interesting new development in Volterra: an anonymous tract had been submitted to the Accademia dei Sepolti on the subject of Allacci and his attack on the Etruscan Antiquities. As Curzio told Holste, who is better known as Lucas Holstenius:

Yesterday in our Academy of the Sepolti, in the urn where we normally put our compositions, a bundle of printed letters was found about the book of Signor Leone Allacci, and I managed to obtain some; I am sending one of them to Your Lordship in order to take this new opportunity to send you my greetings. To my Father, who hoped to have a certain promise from you that you would be coming here, your being held back was a great disappointment, and it also displeased these Signori who greatly desired to see you: and I do not deny that I have not returned home [from Florence] with my purpose accomplished, given that he was not able to receive this favor. Still, I rely on Your Lordship's courtesy to gain satisfaction another time, and in the meantime remind you that I am your most devoted servant and with all my heart I kiss your hands and do you reverence.[1]

The composition was in fact a printed tract, published in Florence by the firm of Amadore Massi, the same printer who had first published Curzio's *Ethruscarum Antiquitatum Fragmenta*. Its author identified himself only by his academic name as a member of the Sepolti of Volterra: Spento, "the Extinguished."[2]

The assignment of such academic names as Spento's constituted one of the major events in the life of a gentlemen's academy. The names themselves were governed by strict standards of thematic appropriateness, debated with elaborate rhetoric pro and con, and finally adopted with immense pomp; the records of the Accademia della Crusca, for example, preserve transcripts of many such exercises. When the Sepolti of Volterra chose a name (and choose—the Sepolti still exist), graveyard humor provided the basic theme, much as the names of the Brutti (the Ugly Men) of Florence punned on their bad looks and the Rozzi (the Ruffians) of Siena on their crude charm. By using his academic name for this public defense of Curzio Inghirami, Spento declared his identity to his intimates in Volterra while disguising himself from readers farther afield.

His tract was not exactly anonymous, but neither was it entirely public; it preserved a gentlemanly discretion in hiding its author from uninitiated eyes. Curzio and his neighbors, however, knew exactly who the "Extinguished One" was.

Spento cannot be identified now with complete certainty, because the membership books of the Sepolti for precisely this period (and only this period) have long been missing from the academy's archives. However, every indication (including the judicious looting of the Sepolti records) points to Curzio Inghirami's best friend, Raffaello Maffei. Maffei's relationship with Curzio mirrored that of his ancestor Mario Maffei with another Inghirami, Tommaso Fedra; Fedra's mother had called Mario Maffei and her son "the twins" despite their eight-year difference in age. More than a century later, Raffaello Maffei and Curzio Inghirami forged the same kind of inseparable bond, although Raffaello was already a married man with children, and, as of 1636, a widower.

Beyond question, however, Spento was a Tuscan gentleman. He professed to have written his twenty-page *Letter about the Book entitled Leone Allacci's Animadversions against the Fragments of Etruscan Antiquities* from the Inghirami villa at Scornello, where he had been introduced both to the scarith and to Allacci's *Animadversiones*, newly arrived from Paris. His essay, as befitted a man of his class, focused above all on his shock at Allacci's bad manners as a polemicist:

Allacci has made use of two techniques in these *Animadversions*: one is to animate his discussion with exclamations, imprecations, and slanders, against Prospero and against Inghirami. Secondly, in order to display his erudition he has on many occasions let the argument wander to matters drawn from his scrapbook in order to fill out the volume, though they are of little or no relevance.

As for the first, I shall only declare it a great disgrace that he should bloody himself and inveigh against the reputation of Prospero, against whom he should contend by means of reasoned argument rather than commonplaces of biting wit and criticism. But I shall declare openly that he offends all good taste when he attacks Inghirami, who is a Cavaliere by birth, a man of blameless manners, and when he began to find these antiquities was so young that he could barely comprehend them, let alone fabricate them. . . . [O]ther *litterati* have not withheld those reservations they thought worthy of consideration in order to arrive at the truth, but they did so within those limits of modesty that become every person, but especially those who make a profession of the arts and letters.[3]

Spento shunned any tangible arguments about the authenticity of the scarith; he sought instead to discredit Allacci largely on the basis of the *Animadversiones'* prose, which was certainly a far less graceful medium than the ornamental Tuscan of Benedetto Buonmattei. Like Buonmattei, he also took an overtly protective tone toward the young antiquarian, another reason to believe that Spento was Raffaello Maffei.

But Spento's good manners and benevolent intentions were not in themselves enough to protect either himself or Curzio Inghirami from continued assaults. In 1642 a Latin treatise began to circulate in Rome under the name of Benno Durkhundurk "the Slav."[4] Its title was almost as big as the book itself: *Benno Durkhundurk the Slav's Examination of the Letter by Spento the Buried Academic on behalf of the Etruscan Antiquities of Inghirami against the Observations of Leone Allacci against the Same.* Within a few pages, Benno succeeded in revealing that his surname, with its rough meaning of "Through-and-through," must have been designed, like his nationality, to suggest plodding stolidity. To his western European reading public, his Slavic nationality would have suggested a tinge of barbarism. The book's false Cologne imprint served notice to the most clever readers that its content was likely to be wicked: ribald, theologically suspect, or satirical.

Benno describes himself as an academic, a resident of Vienna, recently returned from service as a delegate to the Diet of Regensburg, one of the many preliminary meetings between Catholics and Protestants that aimed to put an end to what would eventually be called the Thirty Years' War (Regensburg marked year 23). There Benno claims that he first learned about the Etruscan Antiquities and the controversy that surrounded them, and has now brought home a copy of Spento's letter, which he shares in his confusion with two passing chimney sweeps. Endowed with meaningful classical names—Lignyphagus, "Lamp-eater," and Capnolavus, "Smoke-washer"—these two members of the working class exhibit a degree of classical culture that puts laborious, boring Benno to shame. But in the chimney sweeps' eyes, stolid Benno is pure wit by comparison with Spento, a man so provincial, they observe, that he cannot even write in Latin. All three then lament the fact that a letter composed in Italian remains inaccessible to large parts of the international learned community, suggesting thereby, with no great subtlety, that it is a new barbarism to write in the vulgar tongue.

Furthermore, the chimney sweep Capnolavus turns out to be an Italian, who gleefully attacks Spento's epistolary style as neither idiomatically Tuscan nor elegant vernacular, and, hence, thoroughly unworthy of the title "Etruscan." If Spento will insist upon writing in his regional tongue, Capnolavus sniffs, he might at least do it correctly; this effete Italian workingman turns out to be a linguistic crusader as avid as the members of the Accademia della Crusca. When the two sweeps chime in at the end of the book to give their final judgment on the whole affair of the Etruscan Antiquities, it is with a gentility they fail to observe among their supposed betters; indeed Lignyphagus finally suggests that Capnolavus ought to be elected to the Academy of Chimney Sweeps for his impressive erudition. As if to prove his worth, Capnolavus sums up the situation with masterful authority:

"The ignorance is equal in each of them . . . although it is greater in [Spento] if he does not know that he is dealing with fakes; he knows little or no Latin, and as for his Tuscan, which he knows badly, it is doubtful whether he will use it after this. Rather, I believe that he will keep an honorable silence, and if he gets the itch to reply, he will do it in the Etruscan of the Etruscan Antiquities, so ancient that it goes back to Noah, working the language and his pen in great mystery, spinning tales, mark my words, rather than recounting history. Still, Inghirami is fortunate in having found such a friend, who has invented a new kind of literature to go with these new histories, without a single real fact. Really, it behooves them both not to try to write either in Latin or Italian, but rather in the primeval Etruscan language, which they can fake, and there will be fewer spectators to laugh at them in this theatre of the World." "As for the rest," said Lignyphagus, "Spento should meet the same fate of that monkey who pretended to be an Athenian; for that, a dolphin drowned him, and let it be a timely warning about this forgery of the Etruscan Antiquities, in case any more imitators want to ape it."[5]

The original manuscript of Benno Durkhundurk's *Examination* still exists in the Vatican Library, as part of the collection that once belonged to Cardinal Francesco Barberini.[6] Its handwriting reveals that Benno's original creator was a Hungarian Jesuit, Melchior Inchofer (1585–1649), whose own life had already involved colorful interactions with forgeries, pseudonymous literature, and the Inquisition.[7] Converted to Catholicism in Hungary, Inchofer was educated in Rome at the Jesuits' Collegio Romano, and then posted in 1616 to the Jesuit College in Messina, where he served until 1629 as a lecturer in logic, physics, mathematics, philosophy, and theology. In 1629, however, he published a pamphlet defending the authenticity of Messina's most revered document, a letter allegedly written by the Virgin Mary to the city in 63 A.D. and soon Inchofer discovered, like Curzio Inghirami, that what pleased his neighbors did not necessarily please everyone. In particular, Inchofer's claims on behalf of Messina in this work, *Epistolae Beatae Virginis Mariae ad Messanenses Veritas Vindicata* (The Truth of the Letter of the Blessed Virgin to the People of Messina Vindicated), alarmed the Spanish-backed bishop of Palermo, Cardinal Giovanni Doria, who viewed this Jesuit stronghold in eastern Sicily as a potentially subversive outpost of Rome, and Inchofer's tract as part of a larger plan for papal and Jesuit dominion over the island. Sicily did not yet have its present form of mafia, but Inchofer's superiors made it clear to the neophyte author that because of his imprudent book, he was now in danger of losing his life—or facing the Spanish Inquisition.

On orders from the general of his order, Muzio Vitelleschi, Inchofer escaped across the Strait of Messina to Reggio Calabria and proceeded up the Italian peninsula to Rome, where he arrived in 1630. Again on Vitelleschi's advice, he convinced the Roman branch of the Holy Office to accept a revised version of his book, in which he declared the Virgin's letter as probably, rather than definitely, authentic. In these meetings with the Roman inquisitors, as Leone Allacci would report, "he insinuated himself with many Cardinals of the Holy Congregation of the Index, who became fond of him for his discourse and his handling of the business, so that despite Vitelleschi's resistance he was allowed to stay in Rome. . . . [H]e was assigned a position as confessor at the [Jesuit] church of the Gesù, where he carried out his duties with great zeal and charity."[8]

It was through a shared interest in the Virgin's letter that Inchofer had first entered into correspondence with Cardinal Francesco Barberini; Barberini, in turn, involved him as a consultant to examine Galileo's works, which Inchofer was doing before the *Dialogue on the Two Chief World Systems* ever went to press, let alone before the Inquisition.[9] Thus, in addition to belonging to Rome's most powerful religious order, Inchofer had also made influential friendships in Rome, especially in the circle of Cardinal Barberini, where he kept company with Vatican librarian Lucas Holstenius, Leone Allacci, the promising young Sienese prelate Fabio Chigi, and, by 1635, a fellow Jesuit, Athanasius Kircher, newly arrived to take over the chair of astronomy at the Jesuits' Collegio Romano. In addition to serving as a confessor at the Gesù and a consultant to Cardinal Francesco Barberini, Inchofer also served, like Allacci, as a consultant to the Congregation of the Index; once a defendant before the Inquisition's literary censors, Inchofer now became a censor himself.[10]

Meanwhile, the fictitious Benno Durkhundurk's experiences as a delegate to Regensburg were experienced in real life by Inchofer's acquaintance Fabio Chigi. When Chigi arrived in Cologne in 1639 as papal nuncio, he discovered that the scarith were a favored topic of conversation, a subject that had the twin advantages of being both exciting and politically neutral, at least from the German point of view. Despite the scarith's cryptic references to free will and the Great Aesar, the scholars of northern Europe paid no attention to the texts' religious content. The scholars Chigi met in Cologne and later in Münster had simpler concerns: were the Etruscan Antiquities real or forged, and if forged, who was the forger? And secondly, did the Etruscans write on paper or linen? As Chigi discovered, the answers to these questions were largely settled, no matter how eager the debate around them. The scarith were fake, and the Etruscans wrote on linen cloth.[11] Chigi had given the matter a good deal of thought himself, checking some of Prospero's calendrical calculations to establish the foundation date of Siena and comparing the scarith texts with a transcript of genuine Etruscan inscriptions from Chiusi.[12] But if Chigi had ever been taken in by Prospero's tale—and as a Sienese, he would have had every temptation to do so—he had been thoroughly disabused before arriving in Germany by friends like Leone Allacci and Athanasius Kircher.

Allacci may have demolished the Etruscan Antiquities with his popular book, but he was far from done with them. The manuscript draft of Benno Durkhundurk's *Examination*, still preserved in the Vatican Library, reveals that Melchior Inchofer did not create Benno all by himself; he drew many of Benno's arguments from Allacci's notes on *Ethruscarum Antiquitatum Fragmenta*; some of these notes are now bound together with Inchofer's manuscript. Furthermore, the text itself is covered with changes and corrections in Leone Allacci's tiny meticulous hand. This corrected, collective effort is what ultimately went to press, much like the anonymous *Ad antiquitates etruscas* that was jointly written by Paganino Gaudenzio and Heinrich Ernst. The title page of the *Examination* claims that it was printed in 1642 in Cologne, by Georg Genselin; in fact, it was printed in Lyon with the help of the French libertine scholar Gabriel Naudé, another member of Cardinal Barberini's extensive coterie.[13]

Whereas Allacci's corrections to Benno Durkhundurk's *Examination* mostly served to refine the accuracy of Inchofer's arguments, Benno's change of nationality from Saxon to Slav had another purpose: it muted, or deflected, the ad hominem force of one of Inchofer's satirical barbs. Inchofer had originally cast Benno Durkhundurk as a Saxon. Near the roots of the mighty Inghirami family tree that unfolds at the beginning of Curzio's *Ethruscarum Antiquitatum Fragmenta*, there appears one Enno Billing of Lauenburg in Upper Saxony. Like several Volterran families, the Inghirami traced their origins back to the company of German vassals who had come south in the ninth century with Holy Roman Emperor Otto I and settled permanently in Tuscany: this Enno Billing's son, Engram, would eventually be assigned the title of Count of Pomarance, a castle set on a prominent hill across the river Cecina from Scornello. With his Saxon pronunciation softened on the tongues of his Tuscan neighbors, Engram of Pomarance became the very first Inghiramo Inghirami; eight centuries later, Curzio's father still bore the same old Saxon name.

By changing Benno from a putative Inghirami ancestor to an eastern European, Allacci blunted the sting of his coauthor's stiletto wit. Even after Allacci's alterations, Benno Durkhundurk's *Examination* was still a work of biting satire, one that showed how seriously Allacci's *Animadversiones* had altered the whole tone of debate about the Volterran Antiquities, at least outside of Tuscany. As Fabio Chigi had learned in Germany, there was no longer any real question of needing to attack the scarith per se; Inchofer assumed that Allacci had exposed the forgery once and for all. But, like Allacci, Inchofer largely spared Curzio Inghirami himself as target for abuse, reserving his venom instead for Curzio's defender, mocking Spento's insistence on gentlemanly conduct with ungentlemanly wit, playing Benno's dullness against his own flashing wit. Inchofer, like Allacci, had nearly unlimited access to books, in the Vatican Library, the library of the Collegio Romano, and the private library of Cardinal Barberini. His Order gathered in dispatches from every corner of the planet, but the Jesuits also preserved a huge cache of the forbidden books they were permitted, indeed obliged, to read as consultants for the Congregation of the Index. Loyal Spento, whether he was Raffaello Maffei or some other earnest Volterran, thus found himself compelled to defend his friend Curzio against some of the most learned scholars in the world, with little to aid him but Tuscan manners and a deep emotional connection to their own Etruscan heritage. When it came to polemic, Spento would have been hard-pressed in any case to match Melchior Inchofer, a rootless, resentful creature for whom satire eventually became an irresistible instinct: in his last satire, Inchofer was foolhardy enough to take on the Society of Jesus itself, in a scathing little book called *The Monarchy of the Solipsists*. The "Solipsists," unamused, tried him, convicted him, and sentenced him to life imprisonment in a remote Jesuit house where, like many an inconvenient Jesuit in those troubled times, he was quietly assassinated in 1649.[14]

Writing as Benno Durkhundurk, Inchofer aimed high as a satirist; unlike Spento, who wrote to defend the individual person of Curzio Inghirami, the restless Jesuit struck out at an entire social system rather than an isolated trickster. Hence, rather than attacking Curzio Inghirami or Spento his defender, Inchofer's *Examination* takes on the whole ethos of Tuscan gentlemen's academies, an ethos that Inchofer knew from his investigations of Galileo as well as from his other literary contacts in Italy. His own Society of Jesus, with its wandering, stateless religious devotees, stood in pointed contrast to the highly localized gentlemen's academies as focal points for seventeenth-century intellectual life.

A kindred spirit drove both Inchofer's relatively playful satire and the punishment, neither playful nor satiric, that the Holy Office inflicted in 1633 on Galileo Galilei. They both comprised part of a far larger and more ambitious project on the part of Pope Urban VIII, himself a Tuscan, to subsume Tuscan thinkers and their achievements under the dominion of Rome. Dramatically, both the Jesuit's acid attack on a provincial intellectual and the Inquisition's sentence on a thinker of international significance condemned Tuscan thought per se to a subordinate rank in the European intellectual community.

From Palermo in 1629, Cardinal Giovanni Doria had already observed a connection between the spreading power of the Jesuits and Rome's efforts at cultural dominance, and noted in addition that a crucial factor in this connection was the learned cardinal, Francesco Barberini. Barberini guided Leone Allacci in writing his *Apes Urbanae*, whose first draft of 1632 included Galileo among the luminaries of Barberini Rome. (It was corrected to expunge Galileo before publication in 1633.)[15] In effect, Allacci thereby defined Tuscany's greatest cultural figure as a part of Roman, not Tuscan, society. By 1633 the Barberini family, now in the person of Urban VIII himself, demonstrated another kind of dominance over Tuscan thought by condemning Galileo to silence. Paris had become the new sun in their cultural universe, the Paris of Cardinal Richelieu and, as of 1643, his successor, the transplanted Italian Giulio Mazzarino, Cardinal Mazarin.

In Rome, meanwhile, another of Cardinal Barberini's protégés maintained a curious silence of his own regarding the Etruscan Antiquities. Father Athanasius Kircher, professor of mathematics at the Jesuits' Collegio Romano, was also reputed to command twenty-four languages: Greek, Latin, and Hebrew for his theological study; German, French, and Italian to communicate with friends scattered throughout Europe; as well as Coptic, Arabic, Syriac, Chaldaean, and, most remarkable of all, Egyptian hieroglyphs, which he claimed could be deciphered by comparison with Coptic, the liturgical language of Egyptian Christians. If anyone could have read Etruscan, it was he. But Kircher also had a very good reason to suspend judgment on the scarith. Grand Duke Ferdinando had promised him an Arabic type font from Florence free of charge. Two of Kircher's earlier works on Egyptology had been printed in Rome using the crude if serviceable Arabic set that belonged to the Congregation for the Propagation of the Faith, but for his next project he nurtured more ambitious plans. He had already written hundreds of pages, embellished with his own crude drawings stuck to the paper with red sealing wax, with the idea that they would become his definitive work on Egyptian language, culture, and religion. The book's tentative title, *The Egyptian Oedipus*, cast Kircher in the role of the Greek prince who solved the riddle of the Sphinx (while carefully omitting the other, less enviable aspects of Oedipus' life). It was no time to put the eventual beauty of his masterwork in danger by criticizing the scarith. And so, taking advantage (as he often did) of his fey, diffident personality, Kircher managed to keep as hermetically closemouthed as his mascot, the Egyptian god Harpocrates, a divinity traditionally shown as a baby who holds his finger to his lips. Neither, however, did Kircher defend the scarith, and that silence, eloquent enough in its own way, may be one reason that he had to wait so long for his Arabic font from Grand Duke Ferdinando; he would not obtain it until 1650.

Curzio Attacks

1645

L'Inghirami si difende una pessima causa, ma la difende egregiamente.
Inghirami is defending a deplorable cause, but he defends it brilliantly.

GIOVANNI GIROLAMO CARLI OF SIENA, 1721[1]

Curzio Inghirami completed a draft of his new book on the scarith in 1642 and sent it to Florence for examination by the Holy Office, which granted him permission to print. Then, suddenly, publication stopped. A new correspondent had apparently challenged the Etruscan Antiquities and their authenticity. Curzio therefore set to work—once again—refuting this latest set of criticisms. It was no simple task; only three years later, in 1645, would the Florentine printer Amadore Massi, now flanked by an associate, Lorenzo Landi, see the final result to press. In the meantime, Curzio maintained that he had unearthed many more scarith, so that Prospero of Fiesole's buried hoard now numbered, by his count, two hundred and nine capsules of paper documents. Curzio Inghirami had been in the business of promoting the Volterran Antiquities for eleven years; the youthful prodigy was now a mature man of thirty-one.

His *Discourse by Curzio Inghirami about the Objections made to the Tuscan Antiquities* (*Discorso di Curzio Inghirami sopra l'opposizioni fatte all'antichità Etrusche*) ran to more than one thousand pages, this time written in Tuscan vernacular rather than Latin. Leone Allacci's barbs about the style of *Ethruscarum Antiquitatum Fragmenta* had evidently struck home. Yet the new book followed an old medieval structure: like Peter Lombard, Thomas Aquinas, and their fellow Scholastics, Inghirami divided his defense of the Etruscan Antiquities into twelve categories of argument, addressing individual objections under these headings as a form of point and counterpoint. If he was no longer confident that using Latin would give his own words authority, he could still exploit the scholarly authority of this ancient form of debate, one that most of his contemporaries found excruciatingly boring and reserved for ponderous matters like theology and law.

If the form of Curzio's *Discorso* followed hoary old patterns, its content was new. The objections he assembled and refuted came not only from the printed tracts of Leone Allacci, Heinrich Ernst and Paganino Gaudenzio, and glancing mention by other writers, but also from a series of unpublished letters that Curzio had received over the course of a decade; some readers felt the need to voice their opinions but not to broadcast them in print. Usually, like Vincenzo Noghera and Benedetto Buonmattei, these quiet correspondents obeyed a strong code of chivalry in which the obligation to behave properly overruled the obligation to contribute to the public advancement of scholarship. Predictably, perhaps, his Tuscan correspondents were more gentle than those from farther afield, and many of them, like Benedetto Buonmattei, chose to encourage the young man in private rather than to make public pronouncements.

One of these discreet writers, however, had been persuasive enough to stop the Florentine presses in 1642. He was probably the Abbot Secondo Lancellotti of Perugia, whose *Farfalloni de gl'Antichi Historici* (Tall Tales of the Ancient Historians) had appeared in 1636, just before Prospero of Fiesole and the *Ethruscarum Antiquitatum Fragmenta* could have qualified for inclusion among such canards as "That Maecenas lived for three years without sleeping"; "That there were ever Amazons, and that they lived without men"; the possibility that dragons lived in Italy; and the story of Mucius Scaevola burning off his hand before Lars Porsenna—"Now, this," wrote Lancellotti, "really is too much."[2] Several copies of Lancellotti's criticisms of Curzio are preserved in his papers, along with texts from the scarith that best met Lancellotti's objections—most of these new scarith, remarkably, discovered in 1642. Lancellotti, for the record, eventually pronounced himself thoroughly satisfied by Curzio's arguments.

Many of Inghirami's critics proved less malleable than Abbot Secondo Lancellotti, and the opening paragraph of Curzio's *Discorso*, when it was finally published, admitted that the scarith had been subjected to endless controversy:

The providence of that God who knows all and can do all having disposed that in these times the Antiquities of Tuscany should come to light from the viscera of the earth, by means of which [Antiquities] many facts about Ancient centuries are resurrected from the Tomb of oblivion, it seemed to me the gravest of errors not to publish them in the Press. I have therefore satisfied the World's curiosity, and my own obligations; but because they had never been seen before, for the novelty of the case they have caused grave dissensions among Litterati.[3]

Curzio mounted his grand defense of the Etruscan Antiquities on the basis of new documents as well as the original
scarith. In addition to one hundred new scarith he claimed to
have found at Scornello, he had also obtained access, by permission of Grand Duke Ferdinando, to a manuscript history
of Etruria whose author was nearly as notorious as Curzio himself. The manuscript in question, *De Hetruria Regali* (On Royal
Hetruria), was the work of a Scotsman named Thomas Dempster (1579–1625), who had emigrated to Pisa as a professor of
civil law at the university. By February 1619 Dempster reported
to Cardinal Maffeo Barberini (the future Urban VIII): "I already hear that *Hetruria* pleases the censors, and shall be published this summer,"[4] but his optimism turned out to be ill-
placed. A volatile man under any circumstances, Dempster
became more outrageous than usual that spring, when his beautiful, wayward English wife—whose low-cut gown had already
caused a near riot in Paris—took one of Dempster's own students as her lover. The student was English; hence Dempster's
public wrangles with the couple escalated from what Italians
would otherwise have regarded as a simple case of "the horns"
to a diplomatic problem. By June, Grand Duke Cosimo II had
banished Dempster from Tuscany. As the fugitive wrote to Cardinal Barberini:

I left my library to the creditors, and my *Royal Hetruria*, forced to hand
it over for two hundred pieces of silver, my house confiscated to avoid
jail; attacked on the road by English thugs, I avenged myself at sword-
point.[5]

Dempster's road led him north, over the mountains into the
Papal States, where, with the help of Cardinals Barberini and
Capponi, he obtained a chair in humane letters at the University of Bologna. Grand Duke Cosimo, meanwhile, took possession of *De Hetruria Regali*. In the early eighteenth century,

Anton Maria Salvini, professor at the University of Florence and head of the Accademia della Crusca—in short, the Benedetto Buonmattei of eighteenth-century Florence— passed the manuscript to a rich Englishman, Thomas Coke, Earl of Leicester, who arranged its publication in print during a stay in Florence. Coke's Florentine friend, Filippo Buonarroti, bolstered Dempster's original text with extensive revisions and notes, and the grand duke's own engravers prepared dozens of illustrations of Etruscan artifacts from Tuscan collections. The expanded, updated *De Etruria Regali* (no longer *Hetruria*) emerged from the grand duke's own press over the course of 1723–26, a little over a century after Dempster first gave his *Hetruria* up for two hundred pieces of silver.[6] This two-volume printed edition may have been an antiquarian book as beautiful as *Ethruscarum Antiquitatum Fragmenta*, but it was far more valuable as a scholarly endeavor, for Buonarroti's careful research provided a solid body of factual information to complement Dempster's own wild flights of fancy. Coke may perhaps be pardoned for taking the manuscript home with him when he returned to England; it has lodged ever since in the library of his extraordinary country residence, Holkham Hall.[7]

From its title alone, the original manuscript of *De Hetruria Regali* suggests what a different work it was from its printed version; Filippo Buonarroti's notes survive in manuscript as well, to show how systematically this illustrious scion of Michelangelo's family compared Dempster's treatise with the Etruscan artifacts and historical records available in his own day, quietly but relentlessly revising *De Hetruria Regali* to make it, as best he could, a respectable antiquarian study.[8] Along with his own carefully disciplined reading of ancient authors, he could also add his analysis of a large number of real Etruscan artifacts to Dempster's fantastications. In Dempster's defense, the repertory of Etruscan objects and inscriptions available in the early seventeenth century was still severely limited. The rapid increase in knowledge about the Etruscans during the next hundred years was largely sparked by the discovery of the scarith, and especially by the discussions they generated among "these Signori the Critics."9

Access to Dempster's manuscript helped Curzio Inghirami chiefly in his analysis of ancient historical texts. As the title *De Hetruria Regali* suggests, the Scotsman believed firmly in a bygone Etruscan kingdom, which he described in glowing detail, and this vision of royal Etruria helped Curzio shore up his own invented chronicles of the Etruscan knights and kings (or, as he called them, *larthes* and *lucumones*). Dempster's royal Hetruria may have stretched historical sources as ingeniously as Curzio's Etruscan Antiquities, but Dempster's superior command of Latin and the professorial authority of his writing lent him greater credibility. Curzio absorbed the lesson, presenting himself in his *Discorso* as an experienced Defender of the Etruscan Antiquities rather than a callow, enthusiastic youth, reveling in the patriotism of his Tuscan prose rather than worrying about his Latin. All told, the fat, sturdy *Discorso* exuded an air of dull academic respectability, with its medieval argumentation, its magisterial tone, and its workmanlike quarto format. It could hardly have departed further from the combination of novel and chronicle that had made the splashy folio *Ethruscarum Antiquitatum Fragmenta* a literary event as well as a short-lived antiquarian sensation.

Most readers of Curzio's *Discorso* were looking for something besides the book's claims to respectability. They must have skipped over the hundreds of pages of learned arguments on paper, linen, and details of Etruscan history and cut to the chase: they wanted to know who the forger was. For artifacts that were supposedly genuine, the section of the *Discorso* addressing this question was suspiciously long, but so was the ros-ter of the accused: it included Tommaso Fedra Inghirami, Guillaume Postel, Annius of Viterbo, Thomas Dempster, and, last but certainly not least, Curzio himself. Leone Allacci had already tackled the same issue, although it would have been difficult for Curzio to use the line of argument that Allacci had adopted to exonerate Postel and point the finger in another direction:

Indeed, several people have suspected that the Author of these trifles was Guillaume Postel, a man who was versed in Oriental languages, and gifted at forging history, but This is asserted without reason. To be sure, he was liable to dare anything with great self-assurance, and would indeed do something along these same lines in the Books that he published *On the Region of Etruria*; in which, certainly, he defends the pseudo-Berosus and other fables of Annius, and insinuates many other things that run counter to common opinion on the basis of his intimate knowledge of Oriental languages, but he would never have stooped to the temerity of spreading, with such bold-faced impudence, such half-baked lies; for the prudence of which he drank deep in his readings among the good Authors would have restrained his temerity, and suppressed every license to babble on, and if you compare his unsound opinions and inane proclamations with the ones from Fiesole, good God! How clever they will seem, how prudent, how sober! The former you would say had been written by a man, the latter by a tongueless beast, and a dull-witted one at that. To be sure, Postel was a very bad man, guilty, it is said, of having practiced the evil arts and violated the faith, and condemned by a court of law, and banished, but he never went insane nor ever raved this stupidly and haplessly in his history; however inept he might have been, he would have made these things up—even if they were forged—in accordance with good sense and a greater appearance of authenticity; he would have organized them better, arranged them more aptly, and expressed them more elegantly.[10]

Instead of entering into debates about Postel's and Prospero's literary style, morality, and maturity, Curzio made short work of Postel's candidacy by pointing out that the Frenchman had never come near Volterra, let alone the villa of Scornello, isolated on its remote hill in the Volterran countryside. It was refutation enough.

His discussion of his forebear Tommaso Fedra Inghirami revealed a more complicated mix of sentiments; Curzio, of course, knew Raphael's portrait of this remarkable man directing his wall-eyed gaze heavenward as he waited for inspiration (and probably mourned its sale by the owners in 1640).[11] They even looked alike, with their chubby faces, cleft chins, and alert expressions. Curzio did not want to minimize Fedra's legendary cleverness or his high position in papal Rome, and hence suggested that Inghirami would have made as good a forger as his contemporary Michelangelo, who had achieved some success in the field, at least according to his sixteenth-century biographer, Giorgio Vasari.[12] As Curzio loyally declared:

And if Michelangelo Buonarroti could bury a statue and have it discovered in a short while as if it were antique to display his marvelous talent, how much easier it would have been for Fedra, the Librarian of the Vatican.[13]

On the other hand, Curzio noted that as Vatican librarian, Fedra must have had better ways to spend his time than in forging scarith:

As for saying that Fedra made forgeries in order to emulate Friar Annius ... one answers that he was not the sort of person to aspire to such emulation.[14]

Ironically, this was a point he took from Leone Allacci, who had said of Fedra:

Because this man was extremely intelligent and versed in good literature ... he would never have turned his mind to such trifles ... nor have
arrived at such unbridled gall as to deceive future generations so
shamelessly.[15]

But the real argument against Fedra's authorship, as with
Guillaume Postel, was a simple one: Tommaso Fedra Inghirami
had never spent time at Scornello either. His mother and four
siblings had fled Volterra when he was two, after his father was
killed in the Florentine siege of 1472. He grew up in Florence
and Rome, first in Palazzo Medici and then in the household of
Cardinal Raffaele Riario, and made his career in the Vatican.
If he came to Volterra at all, it was to the family's palazzo in the
city rather than to its rural farmstead.

Some of Curzio's detractors also proposed Annius of
Viterbo or Thomas Dempster as the forger of the scarith,
struck by similarities between Curzio's texts and the pronouncements of these two not entirely trustworthy antiquarians. Again, though, it was difficult to suppose that either had
ever come near Scornello long enough to check out the terrain
and plant scarith along the road to the river Cecina, in among
the tree roots and buried masonry. Curzio absolved them with
alacrity.

Indeed, as Allacci had already noted, there was only one potential forger whose familiarity with the site of Scornello could
be established with ironclad certainty, and that was the Defender of the Etruscan Antiquities himself. Allacci had taken
masterful advantage of the fact, protesting that a man of
Curzio's breeding and erudition would never have stooped to
an act so base as forgery, let alone a forgery as crude as the
scarith of Scornello, and then accusing him anyway in so many
words:

Nonetheless, the spirit senses that Curzio Inghirami, who published similar Fragments and was fortunate in having found them, is a cautious, prudent man, a student of true rather than false and shameless antiquity.[16]

Curzio in turn tried to deflect the force of his familiarity with Scornello by noting that the scarith had been discovered by many people in the course of excavation, not only by himself:

And admittedly, the first Scarith may have been found by my thirteen-year-old sister and myself, but the others have been sought, cleared and excavated by a great quantity and variety of persons, as I have said on other occasions, and if they have not been found where Prospero was under siege, it was nonetheless all on their property. . . .[17]

He cited the police investigation as proof of the excavations' objectivity:

But it can be suspected even less that these writings have been forged in modern times by their publisher, because it is evident that when they first began to be found, the Criminal Court intervened from the very beginning, and then the whole City of Volterra; many people have intervened from almost every place in Tuscany, and all the principal Cities, and places of Italy, and in effect all those who have fancied clarifying their ideas about the truth have been present at the finding of these memorials, and this has not happened on one single occasion, but in a lapse of time, bit by bit, so that from 25 November 1634 up until now these have always been found, and are still continually being found. In addition to this, they are reinforced by the authentic testimony of a public trial, held with the greatest diligence by Signor Tommaso Medici and Signor Ottaviano Capponi, delegated to this task by the Most Serene Grand Duke of Tuscany, in the presence of a nearly infinite number of eyewitnesses, and ratified thereafter by the order of His Highness, who sent the Signori Mario Guiducci and Niccolò Arrighetti, Gentlemen of Florence, to examine the place and watch the excavations, and this document is preserved in the Archive of the City of Volterra.[18]

Finally, however, just as in his defense of Fedra Inghirami, Curzio pleaded that a gentleman would never stoop to counterfeiting. Leone Allacci had made the same contention with pointed irony, but Curzio detailed, firsthand, all the pain, expense, and controversy the Etruscan Antiquities had brought into his life over the course of the previous decade:

In others the purpose of such forgery and imposture cannot be found, and still less can such a purpose be found in me, for I have no other profit than to have spent hundreds of scudi, and no other honor than to have found buried antiquities that could have been found in just the same way by some base plowman; were I to have undertaken such effort, at least I would have wanted to benefit someone by it. But when would it be possible to suspect that I could have done this for some vested interest, once the texts found up to April 1636 had been published, and everything that could have been desired had been accomplished; what purpose, then, after the publication of those, could there have been in fabricating as many more, with the same effort and expense? And all the more seeing that they have had so much resistance and so much opposition? . . . And finally, conceding everything else, given that I am by birth and profession a man of honest conduct, one cannot believe nor assume that I, for any interest or purpose, should have created forgeries or any other such unworthy thing, and no one can believe otherwise who would measure others by his own measure.[19]

For all its epic bulk, however, Curzio's thousand-page *Discorso* did nothing to change the basic terms of debate about the Volterran Antiquities: he had long ago been judged erudite and tempered in his responses, a model of virtuoso decorum, and if his defender Spento had been subjected to merciless dissection of a deficient Tuscan style, Curzio was spared. Style he had, and in profusion. In the early days of the controversy, Benedetto Buonmattei had recommended that "the less you respond to these and any other oppositions made or to be made in this regard, the better it would be,"[20] but in a certain sense Curzio's sturdy volume, by its sheer bulk, put a definitive lid on discussion.

Only one more important sally was fired from Rome, in October 1650. After years of haggling with Grand Duke Ferdinando, Father Athanasius Kircher had finally acquired his fine new Arabic typeface, just in time to display it for the pilgrims thronging to Rome to celebrate the Jubilee of 1650. He had not quite finished his huge and ever-expanding *Egyptian Oedipus*, but he had other temptations to show the more learned visitors among the pilgrim crowds. Most conspicuously, Kircher had been busily at work with the sculptor Gianlorenzo Bernini designing a new fountain for the center of Piazza Navona in Rome, a travertine extravaganza topped by an ancient Egyptian obelisk; Bernini had rescued the broken granite needle by papal order from the ruins of the Circus of Maxentius on the Appian Way. Their spectacular Fountain of the Four Rivers would not be inaugurated until 1651, but Kircher was ready to celebrate it for the Jubilee in a book that was both attractive and useful: for it flattered the pope, Innocent X Pamphili, and attracted future readers for his *Egyptian Oedipus* by means of tantalizing previews. At five hundred pages, it was a slim volume by Kircher's standards—almost a pamphlet. He called it *Obeliscus Pamphilius*, named in honor of Pope Innocent and his family.[21]

Obeliscus Pamphilius devoted most of its pages to an interpretation of the resurrected obelisk of Piazza Navona, referring its hieroglyphic inscriptions—all duly "translated" by Kircher more than a hundred years before the discovery of the Rosetta stone—not only to works of Greek, Latin, Coptic, and Hebrew Scripture, but also to Talmudic and Arabic commentaries. These Arabic citations gave Kircher his excuse, at last, to show off the crystalline precision of his new Florentine font, and he used it profusely. His affairs with the grand duke now settled to his satisfaction, Kircher also decided to settle the matter of his opinion about the Etruscan Antiquities. The verdict takes up one sentence in his 500-page treatise, an indication of how quickly the level of interest in the scarith had fallen since the early 1640s:

The interpretations . . . in Annius of Viterbo's apocryphal Berosus I regard as unworthy, because they have been counterfeited by the Author, and bringing them into discussion is a waste of time and paper. The Volterran's interpretation of the Hetruscan writings he discovered belongs to the same category, which has never brought any benefit hitherto.[22]

With Jesuitical precision, Kircher stated that Curzio's interpretation, not the scarith themselves, was unworthy and counterfeited by the author, but every reader could tell what he really meant. Politics had played a significant part in deflating Curzio Inghirami's audience between 1645 and 1650. The scarith had enjoyed their greatest international fame during the negotiations that led to the end of the Thirty Years' War, when they supplied an attractive yet not-too-heated topic for discussion among the delegates to the series of diets, councils, and meetings that led to the final drafting of the Peace of Westphalia in 1648. With the war over, however, everyone could complain freely about the Peace, and they did. The scarith reverted to a topic of primarily local interest.

Local Tuscan antiquarians persisted in stubbornly defending the authenticity of the Etruscan Antiquities for many years thereafter. Most other scholars by 1648 had expressed skepticism if not outright disdain for what were widely regarded as forgeries, rehearsing the same questions about the scarith's composition and content while adding little of substance to the debate; like the Thirty Years' War, it was over. If the scarith had once propelled Curzio Inghirami beyond the pretty hill of Scornello into the cosmopolitan world, they also caught him in a permanent trap. But when Curzio said that the scarith had brought him nothing but work and expense, he knew that he was telling only part of the truth. They had also brought him his heart's desire.

A Forger's Reasons

1640s

. . . e l'erudito Curzio, che non disdegnò il falso pur di rendere più illustre la sua città. . . .
. . . and the erudite Curzio, who did not disdain forgery so long as
he could make his city more illustrious. . . .

FRANCO PORRETTI, "PALAZZO INGHIRAMI APRE LE SUE PORTE,"
LA NAZIONE (AUGUST 2000)

Leone Allacci obviously believed that the forger of the scarith of Scornello was Curzio Inghirami himself. As he argued in his *Animadversiones*, the sequential discovery of their texts too neatly paralleled Prospero of Fiesole's unfolding tale; another forger, a Tommaso Fedra or a Guillaume Postel, could never have guaranteed that the scarith have been found in so specific, yet so apparently casual an order. Corresponding with Curzio would have given Allacci a simpler means to prove the forger's identity: the handwriting of Curzio Inghirami and Prospero of Fiesole were virtually identical.

The original manuscript of *Ethruscarum Antiquitatum Fragmenta* confirms that Curzio Inghirami wrote the whole book, both the scarith texts and their explanation, in his own distinctive hand, with its lines of spidery script sloping gradually upward from left to right. Leone Allacci was the first reader to call attention to the fact that the scarith's Etruscan texts were written in the same direction as their Latin translations, by noting that both bodies of text were aligned on the left and ragged on the right. Both also had the same slight upward slope as Curzio's own handwriting. This orientation was unusual for Etruscan; in fact, Annius of Viterbo had used the right-to-left direction of the script as one of his chief proofs that the language derived from Hebrew. Allacci was savage in exposing the oversight. He made his argument comparing the scarith texts with a set of fourteen ancient bronze ritual tablets that had been found in Gubbio in the fifteenth century.[1] As we now know, the Gubbio tablets were written in Umbrian, a language related to Latin, rather than Etruscan, which is not. They did, however, use the Etruscan alphabet in its usual orientation from right to left, and no one in the seventeenth century yet knew the difference between Etruscan and Umbrian. Allacci also noted that Curzio's Etruscan letters exhibited so wide a variety that they could never have functioned efficiently as an alphabet of any kind; so he denounced them as pure fiction, and laughably bad fiction at that:

In the Gubbio tablets despite the fact that all the letters are retrograde, they are still Roman letters, and mostly similar to our own, as will be plain to anyone who examines them. But the ones the man from Fiesole writes, although they are frequently similar, still, if they are considered carefully, they differ greatly among one another, and, a matter of no small consequence, the Gubbio tablets are written from right to left, like the Hebrews; the Fiesole texts lead from left to right in the manner of the Greeks and Romans. . . . Observe, then, that in these Hetruscan texts [the "H" is Allacci's clever way of suggesting that the texts are not Etruscan] they are always aligned on the left, and proceed from the same point, whereas they are not [even] on the right, and by this plain argument they begin from no point except from the left. . . .

And so that this [Prospero] will put an end to his babbling, look, where there are notes put before his poems, sometimes repeatedly, by which we are to note the names of the speakers, they are all on the left. O preposterous wit of the man of Fiesole! If the antiquity of the Gubbio Tablets is certain, then there are those who can laugh at [Prospero], and repress the trifling of such an impostor.[2]

In his *Discorso* of 1645, Curzio had asked with considerable rhetorical emphasis why anyone, let alone he, would have done so expensive and strange a thing as forging Etruscan texts. In one of his other works, he supplied the answer. The introduction to Curzio Inghirami's unpublished manuscript *Annals of Tuscany* explains that he had always wanted to become a historian rather than, as his parents hoped, a lawyer. But the writing of history, as he wistfully observed, required a more thorough academic preparation than he had received in the hinterlands of Volterra. Regretfully, he acknowledged the limits that had been imposed on him by his provincial education, but also, as he frankly admitted, his own laziness. His forebear Tommaso Fedra may have grown up as part of the Medici family in Florence, where Lorenzo il Magnifico had engaged the brilliant writer Angelo Poliziano to teach his own sons and Tommaso Fedra had partaken of the same deeply learned environment. But Curzio and Lucrezia Inghirami, in their isolated country villa, took their lessons from Father Vadorini and conversed with their family. However widely they read, their opportunities for further interchange were inevitably limited by the remoteness of their home. In prose as ornamental as Benedetto Buonmattei had taught him to make it, he apologized to his readers:

I so inclined to such a noble subject for study [as history] that it seemed laborious for me to pay attention to Jurisprudence, and I applied my mind with gusto to the ancient records that are preserved in my native city; but I soon realized from the experience that it was no easy matter to write history, and to do so well enough to earn praise for the effort; I knew that I was not endowed by nature with those gifts that would admit me into the company of historians, as I had not cultivated them with sufficient industry, and with the same natural inclination that led me voluntarily to undertake that most noble profession of historian, I incurred the evident risk of being reproved by the world as arrogant. I well recognized that I lacked the study of eloquence, that necessary element of style, nor had I acquired any habit of civil prudence, but for a sin of wit alone, more of conduct than of study, I had passed, if not wasted my time in researching the Archives of Volterra to trace the genealogies of the noble families rather than the Events of history. Inasmuch as I withdrew from the work, I abandoned the enterprise in such a way that even the notes I had made and then neglected were entirely lost.[3]

The discovery of the scarith, of course, changed Curzio's whole plan of study; he could finally put aside his law books and concentrate on sharpening his grasp of Etruscan and Roman history:

Whereupon I will not say that the subject, but rather the goodness and providence of that God who rules and gently disposes all things, did please to make me discover by chance, in the form and manner proven and demonstrated by me elsewhere, what are no longer the fragments, but I may now say the very Etruscan Antiquities, which, though they have not yet ceased to be excavated and extracted from the viscera of the earth, have already raised my spirits; the opposition made to them has forced me to stake my own reputation on my reply, and has given me occasion to consult an author or two and from him learn not a little new information about Tuscan history.[4]

A career in law would eventually have taken Curzio out of Scornello to enroll in one of the Tuscan universities, either Florence or Pisa; his position as Defender of the Etruscan Antiquities kept him close to home except for his public presentations. However acutely he may have sensed the limits of his horizons, educational and social, in Volterra, his writings always showed a deep attachment to the city, to its immemorial history, and to its stark countryside. The scarith texts made the Etruscan world revolve around his home, and explaining them made Curzio Inghirami Volterra's historian:

Adding to this the consideration that Volterra, a City which by common opinion of both Classical and Modern writers has held first place among the principal cities, not only of Tuscany but indeed of Italy, had not had anyone who wrote expressly about her, or at least anyone who had published a History of her. And the gullibility of that Friar, Fra Mario Giovanelli, who dared publish a few things . . . and adding, for the rest of it, what are more the tales of old wives than truths extracted from real texts from the Archives, and giving the whole thing the name of *Historical Chronicle of the Antiquity and Nobility of Volterra*, in my opinion, has rather detracted from the dignity of that city, inasmuch as he made these things appear to have the authority of Great Authors, and Original Documents, and they have once again so tempted me to retrace their history that once again I set myself to work.[5]

And set himself to work he did. Curzio's *Annals of Tuscany* claims to be based on documents from the city archives. A search for these documents, however, will prove frustrating, for many of them have disappeared. Many survive only in a transcription dated 1562 that, like the scarith texts, bears unmistakable signs of Curzio Inghirami's handwriting.[6] The buried scrolls of Prospero of Fiesole, therefore, were only the first and most famous of Curzio's creations. His repertory of falsified papers eventually extended from the last days of the Roman Republic, through the Middle Ages, and into the city's recent history. Forgery had become an impulse as irresistible as anonymous satire became for Melchior Inchofer. Fortunately, the Volterrans proved more indulgent of Curzio's pastime than the Jesuits with their resident satirist—but then Curzio's discoveries all brought luster, however fictitious, to Volterra's place in history.

If the scarith spared Curzio Inghirami a career at law and set him on a historian's (as well as a forger's) path, Cavaliere Giulio took advantage of the Etruscan Antiquities for other reasons; in 1636 he launched a series of lawsuits that were designed to keep the villa and its properties entirely in the Inghirami's male line.[7] His was the kind of litigation that happened frequently in early modern Italy, where inheritance law divided real property equally among descendants rather than passing parcels intact to a single heir by primogeniture. One part of the villa's vast properties had belonged to Father Luigi Inghirami, a canon of Volterra's cathedral who was also the last direct descendant of the Inghirami line that included Tommaso Fedra and his brother Nello. Canon Luigi, as a celibate priest, willed his portion of Scornello to his brother's widow, Verginia Guidi, along with a pair of oil paintings executed for Tommaso Fedra in 1510 by his friend Raphael.[8] One of these was the sublime portrait of the fat, wall-eyed Fedra rapt in thought as his pen pauses over a blank page (now in the Pitti Palace); the other was the lost *Madonna of the Veil* (one copy is kept in the Getty Museum in Los Angeles).[9] When Verginia Guidi remarried, Cavaliere Giulio saw his chance. He sued both for the land and the paintings. He obtained the land.

By 1636 Scornello had became a particularly attractive property, for the Grand Duke of Tuscany decided in that year to make the salt springs of Volterra his prime source for that important commodity. With its commanding view over the valley of Le Moie and its ready access to the salt beds, the villa was an ideal site for placing an official overseer for the salt industry, and this is what Inghiramo Inghirami became by a state decree of 1636.[10] As for Raphael's paintings, the one that certainly survives, the portrait of Tommaso Fedra, still commands the room in which it hangs with its luminous rendering of the great man's pale skin and bright eyes. It entered the Medici collection in 1640, when Verginia Guidi and her second husband sold it to cover one of her gambling debts and the price of litigation with Cavaliere Giulio.[11]

Leone Allacci and his publisher Sebastien Cramoisy had suggested in 1640 that fanatical patriotism might have provided the driving motive for the "Deceiver" who forged the scarith of Scornello, just as greed for personal and patriotic fame, rather than greed for lucre, had driven Annius of Viterbo.[12] With Annius, of course, that insidious, abstract hunger had notoriously driven him from forgery to insanity. The greed in Curzio's case, on the other hand, seemed to have motivated his relatives rather than himself: greed for the property of Scornello (and for two Raphael paintings), combined with greed for vicarious fame in the person of a promising young man who had stubbornly refused to apply himself to the study of law and thus ensure his ascent up the Tuscan social ladder. When he wrote his *Animadversiones*, Allacci could not have known the subtleties of Curzio's situation or of Curzio's mind. For all his callow inexperience and spotty education, Curzio Inghirami was no fool, and at some point his protectors in Tuscany seemed to have recognized this aspect of his character.

As a forger, Curzio Inghirami shows certain parallels with another famous teenaged forger, Thomas Chatterton (1752–1770) of Bristol, who, among other spurious texts, created a series of old English poems in the eighteenth century under the name of Thomas Rowley.[13] Like Curzio, Chatterton did not command enough Latin to pursue a proper academic career; his grammar school education left him as hopelessly provincial as he was ambitious and talented. Ambition brought him to London, where he hoped to make his way as a professional writer, but in that huge, harsh city he could not earn enough to eat on his writing; slow hunger turned quickly to despair. In 1770, at the age of only seventeen, Chatterton swallowed an overdose of arsenic and opium in his London garret to avoid starving to death in public. For both Thomas Chatterton and Curzio Inghirami, forgery was an aggressive act, a way for a young man of restless intelligence to take quiet revenge on a stifling social system, especially on its scholarly elite from which they were so definitely excluded. At the same time, however, their forgeries were also massively comprehensive works of fiction, the expression of genuine imaginative inspiration. For them, no less than for more recent artistic forgers like Alceo Dossena and Eric Hebborn, sincere desire to create drove them as hard as their wicked delight in hoodwinking a gullible—and often pretentious—public.[14]

Curzio's case also differs significantly from Chatterton's. Above all, the extraordinary strength of the family in Tuscan society ensured that Inghirami would live out the whole of his life in a dense social network. He could never have died alone and forgotten in a garret, and his portrait shows that he ran no risk whatsoever of starving to death; the huge, lush family tree he printed as an indispensable part of *Ethruscarum Antiquitatum Fragmenta* gave him both strength and protection. As a result, his forgeries themselves are altogether more social than Chatterton's, providing a contribution to Tuscan civic history rather than chronicling the outpourings of one isolated soul like that of the fictitious Thomas Rowley. Prospero of Fiesole, after all, consigned not only his biography to the scarith, but also, and more importantly, the records of his whole culture.

At the same time, his family's very protection helped to trap Curzio within his own fictions. Once the scarith had been made public, whether they were entirely convincing or not, the family's own *bella figura* required insisting on their authenticity. A young man's unusual pastime determined the plan for the rest of his life.

Curzio's exceptional career may have made him rich in adventure and works of art, but his bold choices had other consequences as well, both social and economic. He married a young Volterran lady, Orsola Ciupi, in 1650, and produced a son, Lino, and two daughters, Silvia and Caterina, in rapid succession.[15] He had already succeeded his father as salt inspector upon Inghiramo's death in 1640 and took an active part in administration of Le Moie on behalf of the Grand Duchy from his vantage in Scornello.[16]

Extraction of salt in Volterra, however, was a complicated and expensive process. The underground beds were penetrated by springs that brought the salt to the earth's surface, where it could be isolated by an elaborate process of drying. In 1650 the grand duke debated closing down the Volterran works and instead buying all of Tuscany's salt from the seaside flats of Trapani on the Sicilian coast.[17] The chief opponents to this move were Curzio Inghirami and Raffaello Maffei. Maffei by then held the position of Provveditore del Sale (Overseer of Salt) for the Grand Duchy; in this capacity he drafted a long treatise about Le Moie (*Trattato delle Moie*), presenting a learned case for their necessity and efficiency.[18] Curzio, for his part, traveled to Livorno in February 1650 to present the case for

Volterra before the grand duke himself. The two friends, now important civic officials, managed in the end to preserve the industry and its revenues for their city, an arrangement that would continue into the mid-twentieth century.[19] On the basis of his long experience in Volterra's archives, Curzio also proposed a thoroughgoing reform of the city's statutes, known thereafter as the "Riforma di Curzio."[20] The reform was never implemented: it would have involved changing centuries of entrenched civic bureaucracy. It stood instead as a magnificent dream, a proposal defeated by its excessive good sense.

In 1647 the city council of Volterra resolved to contribute to a new international project devised by the French Jesuit Jean Bolland: a scholarly edition of the lives of the saints with a huge apparatus of learned notes on their relics and their location (the still ongoing *Acta Sanctorum*). As the birthplace of Saint Linus, the second pope, Volterra had a Christian history to be proud of. The task of drafting Volterra's report for Jean Bolland and his team fell to Curzio Inghirami and Raffaello Maffei, whose fame and erudition were warmly noted in their letter of appointment. They completed their *Processo alle Sacre Reliquie* (Trial of the Sacred Relics) in 1649. Some readers objected to what seemed to be far-fetched claims, but then most saintly relics had accrued a set of extravagant legends, from Saint Galgano's sword, still sunk to this day deep in a boulder to the west of Siena, to the volcano-stopping blood of Naples' beloved Saint Januarius—San Gennaro—first put to the test when Vesuvius erupted in 1631.[21] In between these erudite expeditions, Curzio seems to have embodied the perfect image of a jovial country gentlemen, watching the seasons come and go across Scornello's timeless Tuscan landscape.

In spring of 1655, word reached Scornello that Pope Innocent X had died in Rome; the dramatic eighty-day conclave that followed was maneuvered by a group of young cardinals called the Flying Squadron, whose leader, Fabio Chigi, by now a cardinal and the Vatican's secretary of state, finally emerged as Pope Alexander VII.[22] In December of the same year, Curzio Inghirami died at the age of forty-one. Shortly thereafter, his widow, Orsola, married Raffaello Maffei, himself a widower since 1636. Maffei's biographer, another Raffaello, described how rash the move proved to be:

After the death of Curzio Inghirami, Provveditore Maffei married Orsola Ciupi, the widow of the aforementioned. It seems that this marriage did not please his children by his first wife, nor perhaps the Inghirami family. Certainly it was not a good thing to marry the widow of his best friend. From the marriage contract itself it is clear how little the marriage pleased the Maffei. Among the agreements there is the stipulation that Orsola should not live in the Maffei household but rather in that of her children, together with her maternal uncle, Francesco Luzi.[23]

Orsola and the Inghirami children, that is, remained in Scornello. The younger Maffei continued:

No children were born of this second marriage, and peace did not always reign between the two spouses, and the arguments and recriminations took place especially because of the interests involved. Indeed, at the death of the Provveditore, Orsola, who was seventeen years younger than he, flew quickly into a third marriage, and when she died, her children wanted to bury her in the Inghirami family tomb.[24]

However husband and wife may have argued about "interests," they inevitably mingled. Thus many of Curzio Inghirami's papers have ended up in the Maffei archive, and some of his letters to Raffaello Maffei reverted to the archive of the Inghirami; on paper, at least, the two friends, inseparable in life, are inseparable even now.

Another of their joint enterprises came to light only in 1924, to open a whole set of questions that will probably never find a complete answer. In that year, a little article in the new magazine *Rassegna Volterrana* revealed that a notoriously unreliable compendium of medieval documents in the civic library called the *Estratto del Camerotto di Volterra* (Selection from the Chamber of Volterra), and supposedly dated 1562, instead bore the unmistakable handwriting of Curzio Inghirami.[25] Most historians had avoided this transcript of documents from the Volterran archives ever since its "discovery" in the seventeenth century, but not all: in Florence the historian known as Scipione Ammirato the Younger cited its documents for his revision of the elder Scipione Ammirato's famous *Florentine Histories*— Ammirato Junior would have had to scour Volterra's archives in order to realize that not a single deed in the *Estratto* survived in the original, but only in this later transcript—and he may not have noticed that the handwriting of the *Estratto* looked more like Curzio Inghirami's seventeenth-century script than the penmanship of the mid-sixteenth century.[26] But Scipione Ammirato the Younger was not the only historian to fall for Curzio's creations. The most enthusiastic believer of all was Raffaello Maffei, whose *History of Volterra* (*Storia Volterrana*) from the years 962 to 1351 drew liberally from the *Estratto* to create a vivid picture of Volterra in the Middle Ages, and needless to say, a far more important Volterra than other records had ever suggested.[27]

There is no question whose hand drafted the forged Etruscan documents of Scornello or the false extract of medieval documents from the archives of Volterra; it was Curzio's. But Raffaello Maffei openly collaborated with his friend on the *Processo alle Sacre Reliquie*, and he is surely the person who came to Curzio's rescue as Spento, Defender of the Defender of the Etruscan Antiquities. Maffei was also an enthusiastic amateur poet and prolific amateur dramatist. Like thirteen-year-old Lucrezia Inghirami, whose hair must have wrapped many of the earliest scarith pulled from the ground at Scornello, Raffaello may have been a silent but active partner in Curzio Inghirami's elaborate hoaxes, but just how active a partner we may never know for certain.

❧[VIII]❧

The Sublime Art

READER: *This is the exemplar written by Curzio Inghirami, who, if he had abstained from these lies and directed his literary labors toward investigation of the truth, would have obtained as much glory for himself as, with these commentaries, he made a mockery of his good name.*[1] ON THE FLYLEAF OF THE ORIGINAL MANUSCRIPT OF *ETHRUSCARUM ANTIQUITATUM FRAGMENTA*

Curzio Inghirami made no claims about his ability to write history. He described the historian's task as twofold: to write beautifully and to write wisely, and he despaired of succeeding at either mission:

I set myself to work, although I knew that the difficulties put in my way were insuperable and it was impossible for me to arrive at those prerequisites for composing history, either in matters of style, or for the sublime character that history seeks: both for its ornamentations and the varied descriptions for the discourses, the reasons for the actions that are recounted in histories, and the purposes for which the things recounted have been made to happen; and for the sayings and teachings that History should everywhere supply.[2]

The scarith, however, gave Curzio license to write history by a new set of rules. In addition to creating the scarith themselves, therefore, he also invented a new kind of writing to present them to the public. The result, *Ethruscarum Antiquitatum Fragmenta*, was an uneven work in the eyes of his contemporaries, but Curzio, after all, had barely turned twenty-two when the book emerged from Amadore Massi's press in Florence. Some of the scarith's most scathing detractors had to concede that there were sparks of brilliance amid the nonsense.

What made Curzio Inghirami's history exceptional in its own time was its extensive basis in archaeological data, beginning with Volterra's Etruscan walls, the most extensive example of Etruscan architecture to survive from antiquity. Several of the ancient gates were still largely intact, including the Porta all'Arco with its three colossal heads peering down as they had for millennia.[3] Volterrans like Curzio could not help noticing the contrast between these imposing fortifications and the utter absence of legends connected with them. Puny Chiusi could boast the great warlord Lars Porsenna and lend him to nearby Montepulciano as that city's legendary founder. Arezzo had Gaius Cilnius Maecenas, the Etruscan magnate who sponsored poets in the age of Augustus. Rome had her Etruscan kings and their formidable queens as well as the Vibenna brothers, Aulus and Caelius, who had given their name to one of the Seven Hills, the Caelian.[4] The site of Veii was now a haunt for bandits and had been razed at the beginning of Rome's climb to power, but the citizens of Veii still had their moment of recognition in the histories of Livy and Dionysius of Halicarnassus, and above all in the haunting poem of Propertius:

Woe to you, ancient Veii! Once you, too, were a kingdom,
Once your forum, too, boasted a golden throne.
Now within your walls the pipes of a leisurely shepherd
Sing, and among your bones they're harvesting the fields.[5]

And then there was Viterbo, whose Etruscan history had been rewritten at the end of the fifteenth century by the mad Dominican friar Giovanni Nanni. Writing under the Etruscan name of Annius, the preacher had invented whole dynasties of Etruscan heroes, both for his native city and, after a fortuitous encounter with Pope Alexander VI, for the Rome of the Borgias.[6] A generation after his death in 1502, Annius had been generally recognized as a shameless forger, and rumor had it that he had either died in fetters, the sixteenth-century equivalent of a straitjacket, or been poisoned by Cesare Borgia.[7] Still, his inventions had been so persuasive, so cleverly argued, that no one was entirely certain in Curzio Inghirami's day what was true and what was false about them.

Annius of Viterbo had set the model for subsequent forgers by his deft interweaving of real classical lore and pure invention, relying on ancient (or not-so-ancient) manuscripts, inscriptions, and sculpture to corroborate his tales. He made up texts, but he also carved them in stone (soft alabaster rather than intractable marble). He assembled one "ancient Egyptian" sculpture from various bits and pieces of carved marble of varying ages, much as he assembled his fictitious historical documents from bits and pieces of classical and medieval texts.[8] But Annius was also an intent observer of things Etruscan. He improvised translations of genuine Etruscan inscriptions to satisfy his various well-placed patrons, but he also studied these ancient texts so carefully that he became the first person to read Etruscan script with any real success. His fanciful stories about Viterbo's creation from four small cities were based on the real form of that city (and accurately mirrored the creation of ancient Rome from independent settlements on the proverbial Seven Hills). Annius transformed real place-names into fanciful Etruscan equivalents, for he understood, from reading the Bible and the ancient Roman historians, that the name of a place often contained its history.[9] Annius also inserted Viterbo into the general sweep of universal history, beginning with Adam and Eve and culminating in the reign of his most illustrious patron, the Borgia pope Alexander VI. His intertwining of archaeological artifacts, inscriptions, manuscripts, local topography, and well-known ancient sources not only proved difficult to untangle; it also set the standard for all subsequent antiquarian writing, sincere as well as counterfeit.[10]

Curzio Inghirami knew from Annius and his legacy that a respectable history of Etruscan Volterra had to take its place in the grand sweep of universal history. He would have to include local landmarks and names, ancient scripts and artifacts. He needed to make his own account harmonize with—or better, supplement—the accounts of ancient Greek and Roman authors and, if possible, the Bible.[11] And a proper history needed heroes, preferably heroes who could be traced to living Volterrans, like the Etruscan knights who surrounded Lars Porsenna in a fifteenth-century biography drafted for Pope Pius II by a Florentine cleric named Leonardo Dati. In Dati's *Gesta Porsemnae Regis*, the Etruscan warlord behaves more like King Arthur among the Knights of the Round Table than the archaic potentate described by Livy, but then Porsenna's knights are the ancestors of Dati's potential patrons, identified, like the Knights of the Round Table (who were immensely popular with early modern Italian readers), by vivid insignias destined eventually to become the coats of arms of families like the Piccolomini, Tarugi, and Bellarmini (these last, Dati declared, were descended from two handsome Armenians, the "belli Armeni").[12]

Curzio—nurtured like all his contemporaries on the chivalric romances of Matteo Boiardo, Torquato Tasso, and Lodovico Ariosto—had also adopted the code of chivalry to the exigencies of seventeenth-century life; the etiquette that governed learned academies like the Accademia della Crusca, military orders like the Knights of Saint Stephen, and state-sponsored associations like the universities of Florence and Pisa was still an etiquette of chivalry, however significantly the world and chivalry itself might have been transformed in the meantime.

The latter half of *Ethruscarum Antiquitatum Fragmenta* is a long list of Etruscan cities with their reigning lords and vassal knights, each distinguished by an insignia. This catalog shows the influence of medieval chronicles as well as chivalric literature, and like medieval chronicles it is boring, but it serves its designated purpose. Like Annius of Viterbo and Leonardo Dati, but also like the more humble local writers who abounded in Italian communes from the fourteenth century onward, Curzio Inghirami based his text on the inscriptions and coats of arms in Volterra and its environs that survived, sculpted in stone, from earlier eras, and flattered his neighbors for miles around by providing them with ancestors and a detailed history.[13] As a forger, however, Curzio also did his predecessors one better by adding a new kind of data to his body of supporting evidence: archival documents. The Grand Duchy of Tuscany and its important, efficient bureaucracy were as essential to the young man's understanding of the world and its ways as his chivalrous code of ethics. He made that bureaucracy and its paper trail reach back into Etruscan times.

By contrast with its laborious second half, the first half of *Ethruscarum Antiquitatum Fragmenta* is something quite different, and refreshingly unconventional: a story of discovery that is both a seventeenth-century detective story and a historical novel. As Curzio becomes more and more engrossed in excavating for scarith, the story of Prospero of Fiesole takes its own suspenseful shape; the more eagerly he pursues the student augur and takes an interest in him, the more dire Prospero's fate seems to become. Curzio had made his Etruscan hero as coyly elusive as the nymph-heroine who animates—and then abruptly departs—the quintessential antiquarian romance of early modern Italy, the *Hypnerotomachia Poliphili* of 1499.[14]

Ethruscarum Antiquitatum Fragmenta also provided a representative sample of all the kinds of learning that a seventeenth-century Italian reader would expect from an outstanding work of history. Whether by deliberate design or unconscious inclination, Curzio laced his story with information drawn from every aspect of contemporary learning in his own day, excerpted from the books in Inghiramo Inghirami's library or from his own readings elsewhere. Curzio's writing therefore provides an excellent clue to the range of Curzio's reading.

Thus it may have been no accident that a staunch Galilean like Vincenzo Renieri was so attracted to the scarith's astronomical lore; he must have seen the evident parallel that Curzio implied between Etruscan augurs, priests, and thunder-diviners, on the one hand, and the observations of another sharp-eyed Tuscan, Galileo, on the other.[15] In his own way, Curzio had attempted to strengthen Galileo's cause by referring the great scientist's expertise in astronomy to his Etruscan heritage. If any argument would work to sway the opinion of an angry Florentine pope, surely it was the legacy of Etruria. And because the Etruscans were so famously religious, Prospero's documents fortified Roman Catholic theology by assuring posterity that the Etruscans had favored the doctrine of free will and foreseen the reckoning of time from the moment of Christ's birth.

Curzio Inghirami's Etruscan script is an oddity for its time because it deviates so markedly from real Etruscan; by the

1630s a well-established body of a few genuine inscriptions already existed, passed in manuscript copies from scholar to scholar.[16] Volterrans took pride in two particularly famous inscriptions, both of them on funeral monuments that had been exposed in 1499 during excavations in the area of Volterra's Roman theater. One of these was the funerary stele of an Etruscan warrior named Avile Tite (now dated to the sixth century B.C.); the other was scratched across the arm of a marble statue of a mother and child, now known as the *Kourotrophos Maffei*, that once decorated an Etruscan woman's tomb (*Kourotrophos* is ancient Greek for "Child-nurturer").[17] Both of these monuments had been on display in the Palazzo Maffei in Volterra since their discovery in 1499 by the humanist scholar Raffaele Maffei "Il Volterrano"; as the latter-day Raffaello Maffei's closest friend, Curzio Inghirami must have seen them constantly.[18] These two monuments provide the models for many of the letters in Curzio's overly extensive Etruscan alphabet; it is somewhat surprising that the young antiquarian failed to notice, as Leone Allacci was only too glad to inform him, that both these inscriptions were written right to left.

Some of Curzio's Etruscan letters, on the other hand, look more like Scandinavian runes than any Etruscan inscription. Curzio Inghirami probably had the opportunity to learn about runic scripts, if not at home in his family's library at Scornello, then at least in the libraries of Florence and Pisa. The first connection between runes and the Etruscan alphabet had been made more than a hundred years before, by the Swedish scholar Johannes Magnus.[19] In Curzio's own lifetime, the sudden rise of Swedish military power had made Scandinavian culture a topic of urgent interest, and a fascination with runes developed as part of this more general awareness. By the mid-1630s, when Swedish troops had overrun Poland and parts of Germany, several books on runes became available in Italian vernacular translation as well as their original Latin.[20]

The dramatic flair that Curzio displayed in the opening sections of *Ethruscarum Antiquitatum Fragmenta* eventually found more explicit expression in two comedies: *L'Amico Infido* (The Unfaithful Friend) and *Armilla*. These were never published, although the ravaged state of the two manuscripts of *Armilla* may suggest that it had been performed at least once. The comedies are well-crafted works, set along the lines of Plautus and his early modern followers, and, like Plautus, they are genuinely funny. *L'Amico Infido*, in particular, contains a hilarious send-up of the scarith and the kinds of people who could be presumed to study them, personified in a gullible pedant who speaks a crazy mixture of Latin and Tuscan:

Biduo hic manere decrevi to see those beautiful writings, which I hear were just found miraculously in *quadam* Villa hereabouts *propinqua*. I hear that in *his continentur omnia antiquissima memoranda* of all Europe, which *antehac omnibus fuere ignota*.[21]

In other words, *Ethruscarum Antiquitatum Fragmenta* provided not only a detective story, a universal chronicle, a medieval *chanson de geste*, and a cross section of contemporary topics in scholarship; it was also the work of a sharp-witted humorist. Indeed, if Curzio's large compendium of chivalric lore is read not as a serious work, but as a monumental parody, many of its details suddenly fall into place.

The chief object of his attention, not surprisingly, was none other than his greatest predecessor as a forger of Etruscan antiquities, Annius of Viterbo. But Curzio also devoted a good deal of his ingenuity to aping the prophesying abbot Joachim of Fiore. If Prospero of Fiesole's oracles can be read as bombastic, then they can also be read as deliberately comical:

The Wolf is the mother of the Lamb. The Lamb shall love the Dog. A Pig shall come forth from the horde of Pigs and shall devour the work of the Dog.

Cave, cave, cave[22]

As Leone Allacci pointed out, no reader could say that Curzio had never warned them.

Curzio created dozens of these portentous pronouncements, each one susceptible to interpretation by his readers as deep truth, or as sound and fury, signifying nothing:

The vulture hath raised its voice from the face of the Locust. The Locust shall devour Lions. The stones shall sweat in horror.[23]

No wonder Curzio and Lucrezia Inghirami greeted the discovery of the first scarith by doubling over with laughter. What fun they must have had creating them!

Curzio Inghirami's fun with the scarith was fun of a distinctively Tuscan kind. One of the most time-honored forms of Tuscan humor was (and is) the *beffa*, the practical joke, and the more far-fetched and extensive the *beffa*, the better.[24] Boccaccio's *Decameron* is filled with tales of *beffe*, like the lascivious priest who introduces the chaste but gullible Alibech to the world of sex by convincing her to "put his Devil in her Hell," or the clever Florentine who puts a gold piece up his donkey's posterior to make the creature defecate gold in front of a greedy Sienese. From Boccaccio onward, *beffe* form a staple of Tuscan literature. Judged as a classic Tuscan *beffa* rather than an antiquarian find, the scarith of Scornello were nothing short of magnificent: what started as the prank of two giggling teenagers ended up as Europe's antiquarian sensation, with zealous partisans pro and con scattered from the Baltic coast to the Mediterranean fastness of Malta. And like so many *beffe*, once Curzio's forgeries gained momentum, they charged ahead on their own, out of anyone's control. That was the beauty and the danger of the *beffa* as an art form, a true expression of the Tuscan—or better, perhaps, the Etruscan—soul.[25]

What may have alerted readers to Curzio's real cleverness was the latter's treatment of Annius of Viterbo. In many respects, beginning with the spelling of the title of *Ethruscarum Antiquitatum Fragmenta*, Annius of Viterbo had provided Curzio with his chief model as an author and forger, an emulation that Leone Allacci described as lame, the work of a "blundering clown." Yet Curzio's treatment of Annius of Viterbo involved a more complicated operation. As Inghirami threaded his own way through universal history, seeking to carve out a suitably royal place for Volterra, he both criticized Annius and paid him implicit homage; but most of all, however, *Ethruscarum Antiquitatum Fragmenta* treated Annius of Viterbo to a colossal send-up.

On the simplest level, Curzio laid siege to Annius' patriotism by substituting Volterra for Viterbo wherever Annius had made some extravagant historical claim for his native city, and began this operation with the very first scarith text he transcribed in *Ethruscarum Antiquitatum Fragmenta* (supposedly excavated on the nineteenth of September 1635).[26] Annius had claimed, notoriously, that the city known in his day as Viterbo originally consisted of four settlements on four separate hills; these combined under the kingship of Noah/Janus to form a single city called Etruria.[27] From the outset, the scarith texts "proved" instead that this four-part Etruria was really Volterra— or, as Curzio's text put it, "Vulterra, id est Ethruria."[28] The names of Annius' four hills, as Prospero's hoard revealed, were not the four hills of Viterbo, but rather the names of the four primeval Etruscan tribes. Again, as Leone Allacci admitted, no one could say that Curzio had failed to warn them—"*Cave, cave, cave*" stood on the first page of the very first scarith.

At the same time, Curzio also adopted the pseudo-Etruscan terminology that had made Annius so convincing in his own day, playing along with the made-up meanings of Annius' made-up words and adding crazy words of his own. By correctly reading the Etruscan letter theta on an inscription (the epitaph of one Arnth Sauturinies, son of Larth and Fulni), Annius had come up with the word *larth*, now recognized as one of the most common first names for Etruscan males.[29] In the Dominican's more glamorous vision of the Etruscan past, however, *larth* (perhaps because it was truly an Etruscan word, perhaps by its evident—and genuine—analogy with the name of Lars Porsenna) became the highest office in the land of Etruria, the chief *lucumo* of all the *lucumones*, to use the word for Etruscan chieftains that had been preserved as a gloss by Roman historians.[30] Curzio's Etruscan chronologies bristle with these exalted *larthes* (the plural preferred by both Annius and Curzio), but Lars Porsenna, a real-life *larth* if there ever was one, is stripped of his rank as *lucumo* after leading his unsuccessful expedition against Rome (whereas modern scholars believe that Lars Porsenna succeeded in conquering the city and that the tale of his defeat is crude ancient Roman propaganda).[31] Curzio's *larthes* are flanked by the priests known as Sagi, another Annian term, but also—and this touch is Curzio's own—by librarians whose scarith-stuffed workplace, in Curzio's invented term, is called the Gorg.

Both pseudo-history and parody, a work of comedy with serious implications for the future conduct of its author's life, *Ethruscarum Antiquitatum Fragmenta* ultimately raced out of Curzio's grasp, as a work of literature and as a social phenomenon, so that a bright teenager's *beffa* to while away the winter of 1634 took over his life. Curzio had helped to bring this fate on himself by insisting that the scarith should have an objective reality. As a result, he could never reveal his own artistry as a forger: a good forgery was anonymous forgery, and the moment he claimed to be Prospero of Fiesole, the whole deception would fall apart. When (at least according to Giorgio Vasari) Michelangelo forged an ancient statue and passed it off as antique to a client, his confession proved that he carved just as beautifully as the ancients; his work was judged for its artistic rather than its historical value (and for its considerable value as yet another Tuscan *beffa*).[32] Curzio, whose creations aimed for no absolute artistic value, was forced to revel in his *beffa* alone. For this inconvenience as for so much else, he could thank Annius of Viterbo.

For it was Annius who had provided Curzio with most of the standards by which to measure his own achievements in the counterfeiter's game.[33] He had studied these standards with close attention and knew that unless his "discoveries" could stand on their own as objective documents, they could not stand at all. Annius created a fairly small body of forged texts. The bulk of his great book on Hetruscan antiquities was taken up with the commentaries that wove these texts together into a grand and revolutionary history. Critics, including Curzio himself, could hardly avoid noticing the similarity in style between the prose of the ancient scriptures and the friar's own comments on them; they concluded, rightly, that Annius had written everything himself. Curzio therefore did Annius one better by increasing the volume of ancient texts in his treatise and eliminating the incriminating commentaries as much as possible. As it was, Leone Allacci could still detect a general consistency in the prose of Prospero's texts and held it against the Deceiver:

These writers are always flaccid, languid, abrupt, unkempt, stammering—except where chattering takes hold, barbarous; always afraid to make a mistake, nowhere plump with the rich sap of antiquity, of age, nowhere outstanding. . . . The Gods blast you, you greenhorn! This is a plebeian expression, from the dregs of the rabble.[34]

Despite Allacci's invective, Curzio stuck to his point, on which, in fact, they agreed:

It is said against the books by the authors falsified by Annius that these are written in the same style as the Commentaries by Annius himself, which cannot be said about the Tuscan Antiquities that have been discovered, and published without any comment or addition whatsoever.[35]

Furthermore, as Inghirami noted with disdain (and a slight distortion of the truth), the preacher from Viterbo mostly doctored extant sources, whereas the scarith presented entirely new texts.

Annius falsified Authors who were known to have written; hence [any forger of the scarith] was not compelled to invent unknown, imaginary Authors no longer known, for they would have been granted no credit or respect, given that he had no shortage of ancient writers to whom he could have attributed his hoax. Annius translated and commented on what he had invented, and perhaps the real point of his hoax was not so much the fabrications and fallacies written by his Authors as it was the ability to comment upon them after his own fashion, and also to display his erudition, but this person does not so much as add a syllable. Annius published and brought to light what he himself had invented, whereas this person would have had to respond to the hoax on the basis of information that human reason can no longer retrieve.[36]

In short, Curzio concluded:

The mission of Annius was to restore to Viterbo all the greatness and antiquity that had accrued to other Cities, but the mission of the person who forged the Tuscan Antiquities would have been entirely unlike that of Annius, given that in their texts he impartially accords each City its proper praise without ever disagreeing with the Classical Authors. . . . These writings could not have been invented by Fedra, nor by others, to emulate the fictions of Annius.[37]

Besides, unlike some of Annius' more far-fetched claims to have transcribed information from slabs of alabaster that glowed from within, Curzio insisted that the scarith were available for inspection by any and all:

And the originals have been shown and are continually shown to anyone who would like to see them.[38]

However, examining the scarith in Curzio's presence must have been a distracting show all the same; only in 1700, fortyfive years after his death, did someone notice that the paper on which the texts were written bore the watermark of the state paper factory in Colle di Val d'Elsa.[39]

Eppur si muove

1966

Eppur si muove.
The Earth does so move.
APOCRYPHALLY ATTRIBUTED TO GALILEO,
WHO NEVER SAID IT

In 1926 the young archaeologist Ranuccio Bianchi Bandinelli made a survey of the area south of Siena, searching for possible Etruscan sites. Among other likely areas, he noted a prominent hill near the little *castello* of Murlo and its nearby town, Vescovado di Murlo. The summit of this hill, deserted and covered by scrub, offered a commanding view of the surrounding territory, and local farmers reported finding bronze objects as they cultivated the fields on its slopes. Bianchi Bandinelli himself found a handsome Etruscan belt buckle. What particularly intrigued him, however, was the hill's name: Poggio le Civitate (also known as Poggio Civitate), a name that suggested some kind of civic institution, despite the fact that no building of

136

consequence could any longer be seen there. More tantalizingly still, one section of the hill was known as Pian del Tesoro, "Treasure Flats."[1]

Exactly forty years later, in 1966, one of Bianchi Bandinelli's protégés, an American named Kyle Phillips, obtained permission to excavate Poggio Civitate from the superintendent of antiquities for the region of Tuscany. Intrigued by the name of Pian del Tesoro, he sunk his first trenches there to search for its buried treasure.[2] Within a few days, he and his excavators had exposed the corner of a large Etruscan building, which turned out to be the second such structure erected on the site. The first had burned in the seventh century B.C.; its successor, on the other hand, was systematically pulled down and destroyed in about 525 B.C. in what was evidently a ritual demolition. The site was never rebuilt, but it was easy to see why it might have inspired lasting legends. The debris contained the smashed remains of nearly life-size terra-cotta statues that once perched along the building's ridgepoles, peering majestically down into its spacious courtyard.[3] Its roofs had been elaborately decorated in the Etruscan style, with terra-cotta lions' heads and decorative ceramic panels to protect the roof's wooden beams. The ground was littered with badly burnt pieces of richly carved ivory and fine bronze; once there must also have been beautiful works of wood and textile that had decomposed over the course of two millennia.

Surviving records in the Sienese state archives, from 1318 onward, refer only to a few scattered farmhouses on the hill they record as Poggio le Civitate, or "Le Civitate"; one fourteenth-century record mentions an "old, large wall" that may have been Etruscan, but there is no way of knowing for certain.[4] For the bureaucrats of the Sienese city-state, this detail was unimportant, and no one in the fourteenth century but a moody Tuscan transplant to Provence named Francesco Petrarca had yet had much time to spend thinking obsessively about the ancient past. Within a century, of course, that moody Tuscan would have transformed the intellectual life of Europe and brought on a continent-wide passion for antiquarian studies: the figurative rebirth of antiquity that defined the Renaissance. And fifteenth-century Tuscany would rediscover the Etruscans.[5]

Fervent participants in the same passion for antiquity that had stirred Francesco Petrarca in the fourteenth century, both Ranuccio Bianchi Bandinelli and Kyle Phillips were also, like him, unconventional thinkers: Bianchi Bandinelli was both a Sienese count, who could claim a pope among his medieval ancestors, and a committed Communist; Phillips, a shy New Englander who thrived most easily among foreigners.[6] In addition to their stubbornly original patterns of reasoning, they shared deep ties to the countryside that lent them an almost instinctive sense for the places favored by the Etruscans for their tombs, temples, and settlements. Bianchi Bandinelli's Italian colleagues, unlike Bianchi Bandinelli himself (and, exceptionally, the great Volterran archaeologist Enrico Fiumi), had seen no promise in the desolate scrub of Poggio le Civitate; they granted Phillips an excavation permit with a wonderment that he himself summarized as "if the Americans want to dig in the bushes, let them."[7]

Phillips's interpretation of his discovery was equally unusual; he described the building as a temple and meetinghouse that had once served as the ritual center for a league of Etruscan cities.[8] Ancient Greek and Roman authors described one such sanctuary situated north of Rome, the Fanum Voltumnae (Shrine of Voltumna, a deity whose gender is still in doubt).[9] One historian, Dionysius of Halicarnassus, mentioned a second Etruscan league situated to the north; perhaps Poggio le Civitate was its meeting place.[10] In any event, like Bianchi

Bandinelli, Phillips took the place-name "le Civitate" seriously; it was a plural, not a singular—"the Cities," not "the City." For the shrine's destroyer, he proposed Lars Porsenna, bent on concentrating his own authority from Chiusi by destroying other Etruscan centers of power.[11]

Most Italian archaeologists have been unwilling to subscribe to so unusual a view; they have described the building on Poggio le Civitate as a "palazzo" for an Etruscan prince, pulled down in the social upheavals that replaced generations of Etruscan "princes" with rich, ambitious warlords like Lars Porsenna and Thefarie Velianas, the magnate who endowed a religious sanctuary on the Tuscan coast and commemorated the fact in tablets of pure gold.[12] If Phillips's American background seemed to make the civic sanctuary at Poggio le Civitate sound somewhat like the stateless enclave of Washington, D.C., Italian archaeologists' description of the princely palazzo partook more of the world of Lorenzo de' Medici and Machiavelli than that of Lars Porsenna; thus, no differently than in Curzio Inghirami's day, ideas about the Etruscans still tend to reflect the people who study them.[13] In any event, it took two thinkers as independent as Bianchi Bandinelli and Phillips to understand how a combination of strategic site, promising surface finds, and local tradition could combine to point toward an important Etruscan settlement at Poggio le Civitate, even in the absence of substantial architectural remains aboveground.

A Tuscan countryman just like Ranuccio Bianchi Bandinelli, Curzio Inghirami had spent his life traversing the gentle slopes that led from Scornello to Volterra and the salt springs of Le Moie, as well as the more precipitous drop down to the bed of the Torrente Zambra. However fancifully he embroidered his tale of Prospero's citadel, the story itself was based on intimate knowledge of—and affection for—the hill from which, like the

Etruscan student of his fictions, he had looked out upon the world.

Nor was Curzio's knowledge of the Etruscans entirely imaginary. At least one of his purported finds from the "citadel" of Scornello was a genuine bronze fibula that would now be dated to the seventh century B.C.; Curzio reported finding it together with a broken jar.[14] The Iron Age ancestors of the Etruscans cremated their dead and buried them in terra-cotta urns, some round, some shaped like houses, some capped by warriors' helmets, and accompanied by treasures like weapons, jewelry, or the ornamental pins—fibulae—that held up their garments.[15] Curzio misidentified his own fibula as a lamp, but his description of its context in the countryside around Scornello is otherwise entirely consistent with this type of Etruscan object, which almost certainly came from an Iron Age burial. Normally Etruscan tombs are found separated from residential areas by a valley with a stream or river to wash off the taint of death. On high settlements like Volterra and Orvieto, however, where running water is far away, tombs also occur on the lower slopes of the hill that supports the city itself. Curzio's description is not precise enough to pinpoint the exact place where he found the fibula; Volterra's territory, which was more sparsely settled than many regions of Etruria, is riddled with Etruscan tombs, just as it was once peppered with Etruscan farmsteads and cities.[16]

In this Etruscan landscape, however, the hill known as Poggio Scornello is one of the most prominent features to the south of Volterra. The steep southern slope of the hill commands a broad view of the valley formed where the Zambra torrent flows into the river Cecina (with its good Etruscan name), and more specifically exploits a peerless vantage to the west over the salt springs called Le Moie. These unusual formations were only used from the fourteenth century onward, but the other mineral resources of this geologically complex region have been exploited since the Late Stone Age and provided the chief reason for Volterra's prominence not only in the ancient Etruscan economy, but also that of medieval and Renaissance Tuscany.[17] That same mineral wealth also made Volterra's territory a constant object of contention; it is no accident that the city occupies the highest promontory in the whole region, its fierce Etruscan bastions incorporated into the Medici castle that was erected after the Florentine siege of 1472. Both were erected to protect the salt works of Le Moïe and the alum beds that lay nearby; it was greed for salt and alum that led Lorenzo de' Medici to covet Volterra from the outset and led his descendants to keep an Overseer of Salt (Provveditore del Sale) as an agent in place.[18]

With one of Volterra's most precious resources lying literally at its feet, Poggio Scornello is not only a plausible site for a minor citadel; it is an obvious one, in the Iron Age, in Etruscan times, in the Middle Ages, in the years before Tuscany became a single political entity. The modern example of Poggio le Civitate shows that imposing constructions may well lie buried under a cover of Mediterranean scrub. Curzio could see another remarkable overgrown citadel from his own home at Scornello: Monte Voltraio, a steep conical outcrop of rock just east of Volterra, where a castle had first been erected by the German emperor Otto I in the tenth century.[19] The castle was inhabited until the early sixteenth century and then abandoned. By the 1520s, Volterra's annexation by Florence had made an outlying fort unnecessary for half a century; in peacetime, the precipitous climb to its summit became a nuisance rather than an advantage. Now Monte Voltraio looks as if it had always been covered by greenery; it may already have looked much the same by Curzio's time, and yet the castle was once so powerful that it harbored a group of Volterran rebels for years undisturbed.

Like Monte Voltraio, Scornello's impressive hill (fig. 19) is both a likely place for significant architectural remains and a place where architectural remains of some sort evidently existed, their presence confirmed by a credible variety of sources, including one of the state engineers for the Grand Duchy, Francesco Fantoni.[20] The police records in Volterra clearly record the original excavators' discovery of thick masonry walls on Prospero's "citadel" and the confirmation of that discovery by the government engineers. The fact that these structures were stuffed with scarith seemed to prove their antiquity, and the walls and foundations, in turn, seemed to prove the antiquity of the scarith, but of course walls and scarith were unrelated until Curzio Inghirami put them together into a tale of war, loss, a student priest, and an archaeological discovery.

The existence of these walls is not as far-fetched as it may seem from the flat narrative of the seventeenth-century police reports, or from Domenico Vadorini's drawing of the site. Neither kind of record fully conveys the steep pitch of Scornello's slopes; Curzio would not have needed to dig particularly deep to plant his capsules if he dug straight into the side of the hill, and the excavators who report walls buried under two cubits (*braccia*) of soil may not have had to move two *braccia* of soil in order to strike masonry. Even the scarith tucked in among ancient tree roots are plausible enough, because sharp drops and erosion have exposed the roots of trees all along the crest of Poggio Scornello, especially along the old cypress-lined road that leads from Curzio's villa down to the river Cecina. It would not be hard, today or in Curzio's time, to plant scarith among the roots of these ancient trees and make them look as if they had been buried for ages.

19 : View of Scornello from Volterra. Author's photo.

For five centuries at least, Scornello has been sparsely populated; the surrounding soil is fertile only with hard labor and for centuries has supported only scattered families of tenant farmers in a manner that a modern scholar has termed "sober."[21] As the police remarked in 1635:

At present in said location there is nothing but a House for the use of the Master, and another for the use of the Farmers and Sharecroppers ... for one hundred twenty years they have been in their present state, indeed, more forested and wild.[22]

The various records of excavation at Scornello in 1634 and 1635 describe the discoveries with some consistency. The first indications emerged piecemeal, with Curzio's inventions, the scarith and the tin idol, mixed in with a real Etruscan pin, what were probably real walls, and layers of ash that could have been old or new or both or neither:

Whereas on 13 December a more diligent search was made and there was discovered beneath a very strong wall that ran from the South to the east some two *braccia* underground, underneath a great Stone that seemed to be made of Plaster, in between two White stones in the form of an Urn, a larger compound than the fist, on which there were some Etruscan letters which could not be recognized because of their age. . . .[23]

Said place before exploration was begun showed nothing but piles of stones, scrub, and bushes, and walls could not be recognized aboveground except in two or three places.[24]

Whereas on the 29 of the same Month more diligent explorations were made beginning in the South toward the place where the first text was found, and searching in said location large, strong plaster walls were discovered, cracked vaults, and other ruins.[25]

Whereas among the other ruins and ruined foundation Walls there was upright scrub, that is, Oaks, pistachio trees, and bay with long roots that demonstrated great antiquity and also great Trunks, deteriorated because they had been cut repeatedly and not replaced.[26]

Whereas the ruins of broken Roof Tiles, Bricks, and stones and walls, even those of the vault, were placed just as chance had left them, hence from the first ruin onward it did not seem possible that they could have been sought out, excavated, or moved.[27]

Whereas in many places Ashes and carbon were found, although deteriorated, and half-burnt human bones, everything confused among the Ruins. . . .[28]

Whereas on the 30 of December in the morning in the last foundation of a Wall some two *braccia* underground there was found a capsule similar to the Text mentioned [above].[29]

Whereas on the 5 of February seven ancient Roman paces distant from the aforementioned location at the end of said Wall, proceeding westward, there was found in the earth a statuette of tin with Etruscan letters on the hem of its dress, and a Bronze instrument, and above the wall a tin container was built in, in which there . . . was the Text mentioned [above].[30]

Whereas in exploring said place it was noted and observed that there was a Citadel of round form measuring 200 *braccia* and that toward the east it had another building adjoining it.[31]

The Volterran police investigators finally synthesized all this scattered information in a definitive report:

[6 June 1635] [Scornello] was seen to be a hill at the roots of which toward the east is the little river Zambra, which flows into the Cecina, which runs from South to West, and flows into the Sea. From East to West there are several buildings belonging to Le Moie and the Salt works, and from this direction the climb up the Hill is extremely easy, but from the Zambra it is steep and difficult. To the north, about three and a half miles distant, is the City of Volterra, and four hundred *braccia* lower in the same direction is the house of the Villa of Signor Inghiramo Inghirami that is the residence of the master, and the farmers. Near the same place is the street that leads to the river Cecina and to Pomarance. The aforementioned Signori Deputies considered the countryside very carefully, and it was seen to be a place for grazing, and wild, and full of scrub and bushes, and trees, some of which, because they had been uprooted in the process of excavation, showed from their extensive roots and thick trunks that they were born in that place in the most ancient times. In addition, many traces of old walls were seen, of which only a few traces remained aboveground, one to the south facing the river Cecina, and the other to the north, and the said walls were examined by Messer Francesco Fantoni, one of the Engineers of the Magistrate of the Signor Captain of the Party in Florence, and by Messer Giovanni Maria Sanfinochii of Volterra and it came to be recognized that they were within a round circuit of walls of 200 Florentine *braccia*, and that this circuit indicated a Citadel. Some other adjoining walls were also examined and other remains were examined in different places nearby with parts of thick walls, which were seen to be Citadel walls, and it became clear that there had been a Castle of 3207 Florentine *braccia* in circumference. Within the perimeter of these walls many ruins can be discerned, and walls of very ancient time, and therefore it was ordered that Fantoni make a plan with every diligence.[32]

The structures described by the police report seem to be located in the area of Poggio Scornello that is now called Porta Santa, "the Holy Gate."[33] The old road that leads from the villa of Scornello to the "citadel" at Porta Santa and on to the Cecina valley is itself supported by substantial masonry substructures that are still clearly visible today; perhaps Curzio may have incorporated this old roadbed into his vision of Scornello in ancient days. There is no surviving surface evidence now of his circular bastion, the towers, or the ruined vaults; the only conspicuously unusual feature of the terrain is a circular stand of old cypresses that was once used as a trap for birds. Dense scrub covers the crown of Scornello, whose chalky soil is still used for pasture, olives, and grain, just as it was when Domenico Vadorini drew the same countryside in 1636 for *Ethruscarum Antiquitatum Fragmenta* (fig. 15), in a map that combines two bird's-eye views with an accurate representation of land use that are remarkably sophisticated for their time.[34]

To judge from the spare and, to a certain extent, credulous descriptions of the police dossier, the masonry remains described by the granducal investigators are more compatible with a medieval castle (the round bastion) than with an Etruscan building like the structure at Poggio le Civitate. In the Volterran countryside, Poggio Scornello is only slightly less prominent than Monte Voltraio, and because of its proximity to the river and the salt beds, its location must have had comparable strategic significance. The arrival of Holy Roman Emperor Otto in the tenth century led to a proliferation of castles around Volterra, manned by his German vassals, including that reputed Saxon forebear of Curzio Inghirami, Enno Billing of Lauenburg, father of Engram/Inghiramo, Count of Pomarance.

For all the time Curzio Inghirami spent exploring and enhancing the archives of Volterra, he might just as easily have rediscovered one of Emperor Otto's castles on his ancestral property rather than imagining the Etruscan citadel of Prospero of Fiesole. It is certainly possible that he, and the Tuscan government officials after him, came across the remains of a medieval fort; in fact, the villa of Scornello itself is said to incorporate a tower that might also have been part of such a complex.[35]

But because Curzio Inghirami was not interested in medieval castles, he imagined an Etruscan fortress, and the names of the fields around the villa at Scornello are now as suggestive of Etruscan hoards as "Pian del Tesoro" had been to the explorers of Poggio le Civitate, Ranuccio Bianchi Bandinelli and Kyle Phillips. Right where Curzio and Domenico Vadorini placed Prospero's bastions, the modern property records of Volterra place the fields known as Poggio del Tesoro, "Hill of the Treasure"; Poggio al Tesoro, "Treasure Hill"; Poggio ai Guardiani, "Guardians' Hill"; and Porta Santa, "the Holy Gate."[36] Curzio's vision of Scornello as an Etruscan citadel therefore lives on in Volterra's archives, obedient to the instructions from the very first scarith: "You have discovered the treasure. Mark the spot, and go away." Curzio marked the spot. Archaeologists, remembering Curzio and his fakes, have stayed away. Yet Curzio found at least one Etruscan artifact on Poggio Scornello. Like the Pian del Tesoro on Poggio le Civitate, the Etruscan soil of Poggio del Tesoro also yielded up at least one real Etruscan treasure, and almost certainly holds more.

To his credit, like Annius of Viterbo before him, Curzio Inghirami learned a good deal about the Etruscans in the course of making up stories about them. Both forgers, moreover, learned not only by reading ancient writers, but also by examining the material world, both its landscape and its artifacts, and acquiring an intuitive sense for every aspect of the ancient human environment. Annius, indeed, was not only a forger but a pioneering scholar, the first person since antiquity to read Etruscan script with some success, by correctly deciphering the O-shaped Etruscan letter theta to arrive at genuine words like *larth*, the proper name that he misread as a kingly office.[37] Curzio Inghirami was far less gifted than Annius as a linguist, but his methods for identifying a potential Etruscan settlement were no different in substance from those used by modern archaeologists, who survey sites for their topographical features and scan the ground for promising surface finds. Fortunately for Curzio, who admitted a certain degree of laziness among his vices, the hill on which he lived provided an unusually apt location for imagining an Etruscan watchtower. He did not have to walk far to plant his scarith. Indeed, the government investigators noted with wonder how close Prospero's citadel came to the *strada maestra*, the cypress-lined road that led from Scornello down to the river Cecina.[38]

But even some of Curzio's more fanciful reconstructions turn out not to be so fanciful after all. His Etruscan cemeteries are dominated by an architectural form that enjoyed great popularity from the fifteenth century onward, a combination of pyramid and obelisk. Although there was real confusion among early antiquarians about the proper shape of both kinds of ancient Egyptian monuments, this confusion may have its basis in reality. Etruscan cemeteries around the ancient town of Barbarano Romano, just north of Rome, show that the Etruscans actually used this peculiar but graceful hybrid form to mark their tombs, just as Father Vadorini imagined them doing in his drawing of Etruscan Volterra.[39] Four centuries nearer Etruscan times than we, Curzio may have seen the eroded

20 : Etruscan heads, Porta all'Arco, Volterra. Fototeca, American Academy in Rome.

Etruscan faces on Volterra's Porta all'Arco (fig. 20) with much greater clarity than we do in an age of air pollution and acid rain. He may have seen other Etruscan monuments as well— and not all of them of his own making.

The treasures that Curzio collected, both those that he found and those that he made, were eventually transferred to Palazzo Inghirami in Volterra, an imposing complex that sits literally on the city's Etruscan wall. Volterra's Enrico Fiumi Archaeological Park was once the Inghirami family's private garden. Some Volterrans claim to remember seeing a wooden case full of *scaritti* in the family collection, among bronze statuettes, coins, and alabaster sarcophagi, although the present generations of the Inghirami family do not remember it. Nor will it be easy to sharpen their memories. Toward the end of the twentieth century, around the time that the Italian state declared 1985 an official "Year of the Etruscans," thieves broke into Palazzo Inghirami and walked off with every ancient artifact they could carry. Someone, at least, still believed that the scarith were real. No wonder Curzio Inghirami is still smiling at his *beffa*.

─────────────────── ⟨ ✣ ⟩ ───────────────────

Afterword

───────────────────────────────────

I first came across Curzio Inghirami and his Etruscan Antiquities in the "E" section of the card catalog at the Vatican Library. I took a quick look at his big book, *Ethruscarum Antiquitatum Fragmenta*, thought to myself "bogus," and put it away. Eventually I realized that "true or false?" was the wrong question to be asking of this young Tuscan nobleman and the evidently pseudo-Etruscan artifacts he claimed to have found behind his house in 1634 and called "scarith." The controversy they aroused involved Europeans of every nationality, as my friend Thomas Cerbu was discovering in the course of his own work on the Rome-based Greek, Leone Allacci. One evening, after we had both been thinking about the scarith for years, discovered the identity of Benno Durkhundurk, and pieced together the scholarly debates, Tom phoned me and asked, "Do you know what it's really about?" "Galileo," I answered. "Exactly," he replied. We had realized simultaneously that much of Curzio's story had to do with relations between Tuscany and

Rome in the aftermath of the sentence leveled against Galileo, the shining light of Tuscan intellectual life, by the Roman Inquisition in 1633. We still wondered at that point how many objects Curzio had found, and indeed whether they had ever existed. The answers came when I took my first trip to Volterra, and told the librarian, the amazing Angelo Marrucci, that I was working on Curzio Inghirami and his *scaritti*—the word still has meaning in Volterra three and a half centuries after Curzio made it up. With a smile, Marrucci said, "Do you know about the other *falsi*? He didn't stop with the Etruscans." I also read Curzio's papers on that occasion; they proved in turn that his story was more than a matter of Tuscany fighting Rome to save its own face. Curzio himself declared that the discovery of the scarith spared him law school; his notoriety, then, was not entirely the creation of an ambitious uncle, but rather the direct result of his own actions. He was a free spirit in an age of repression. In that same visit to Volterra, I saved reading Curzio's plays for last, suspecting that they would be comedies as bad as the scarith were bad Etruscan. Once again I sold Curzio short: I sat in the library giggling over the antics of his characters and their sharp Tuscan comments, and realized at long last that Curzio Inghirami had been a humorist among his many other talents. Perhaps, then, the pomposity of his ancient Etruscan documents was knowingly pompous, in which case the story of the scarith involved parody as well as forgery, and nothing was quite what it had seemed before.

One important piece of Curzio's puzzle remained, but I did not address it for several years. The scarith were too strange a topic to interest publishers, and I wrote about other subjects, including a character who stood carefully to the side of the scarith affair, Athanasius Kircher—who, like Curzio, had his own good reasons to act as he did. It was a random search on the Internet under the word "Scornello" that introduced me at last to the Inghirami family and to Curzio's home (and, incidentally, to Luc Deitz of the National Library in Luxembourg, whose exhaustive research on Curzio took place in parallel to mine, with remarkably complementary results). Scornello, isolated on its hill in some of the most stark and most beautiful of all Tuscan landscapes, had its own story to tell; so, too, did Iacopo Inghirami, who gave me access to the family archive and palazzo in Volterra, as well as to the villa at Scornello, and who realized, once we had jumped over a barbed-wire fence to explore the site where the scarith were found, that we were in a bullpen and he was wearing a red shirt. It became clear both that Scornello must once have been an Etruscan site of some sort, and that one of Curzio's motives must have been a deep love for that strange spare countryside, one he never left for long. Curzio's wit has always kept one step ahead of his researcher, who only now perhaps can take its measure. Scornello is now a working ranch, reachable only by a dirt road, a gentle presence in a landscape that seems unspoiled but, like most of the Tuscan countryside, results instead from a millennial give-and-take between land and people. May it stay that way forever.

Notes

Unless otherwise noted, all translations are my own.

Chapter One

1. Curzio's birth and death dates (29 December 1614–13 December 1655) are recorded on a document in the unpaginated folder entitled "Archivio Maffei" in Volterra, Biblioteca Comunale Guarnacci.

2. The history of the property of Scornello is traced in Costantino Caciagli, *La casa colonica ed il paesaggio agrario nel volterrano* (Pisa: Bandecchi e Vivaldi, 1989), 110.

3. Curzio Inghirami, *Ethruscarum Antiquitatum Fragmenta* (Florence: Amadore Massi, 1636, with a false imprint of Frankfurt, 1637), †v–†2r: . . . *cum a prandio piscatum iremus die B. Cathareinae [!] Virgini et Martryi dicato, dum longe a nostra Domo cubitos circiter tercentos Cecinam flumen inspicio, famulos expecto, et in declive animi gratia saxa devolvo, accidit, ut altero magno saxo emoto exiguus et subniger globulus. . . . Hic saepius a me coniectus tandem casu effringatur. Tum pilos, sub quibusdam corticibus, quibus ipse globulus compactus erat perspexi: his*

rebus in admirationem versus, ipsum magno cum labore dissolui. Primum corticem ex bitumine, pice, resina, cera, thure, storace, et mastice, aliisque huiusmodi compactum esse multi existimarunt. Secundus, qui validior erat, iisdem pilis immixtus tela vinciebatur, quae in pulverem comminuta est; sub illa quaedam charta lintea hisce characteribus notata apparuit. . . . Ipsa vero aliam, quae ob ictus in frusta abiit, continebat cum hoc vaticinio adamussim expresso.

Anno a Rege Iudaeorum nunciato M.dC.XXIIII. Crucifixo M.D.XCI.

Veniet Canis fideliter serviet servitute libera ix annos, et amplius. Lupa mater agni. Agnus amabit Canem. Veniet Porcus de grege Porcorum, et devorabit labores Canis.

Cave, cave, cave.

Prosperus Fesulanus huius Castri accola, arcis Custos Vaticinatus est anno post Catilinam extinctum.

Thesaurum invenisti, locum signa, et abi.

4. See chapter II.

5. The Cecina family, still active in Volterra, appears on Etruscan sarcophagi as "ceicna," "ceicnai," and "ceicnal"; see Murray Fowler and Richard George Wolfe, *Materials for the Study of the Etruscan Language* (Rome: Edizioni dell'Ateneo, 1980), 287. The modern given name "Aulo" is likewise of Etruscan origin (written Aule; see ibid., 259–63); the eighteenth-century Volterran antiquarian Lorenzo Aulo Cecina bore a name that not only extended back into Etruscan times but was also made famous in the last days of the Roman Republic by Cicero's friend the Volterran augur Aulus Caecina. The Inghirami, on the other hand, traced their origins back to a certain Enno Billing of Lauenburg, who emigrated from Saxony in the tenth century A.D.; see chapters V, IX.

6. Marjorie Reeves, *Joachim of Fiore and the Prophetic Future* (London: SPCK, 1976); Reeves, *The Influence of Prophecy in the Later Middle Ages: A Study in Joachimism* (Oxford: Clarendon Press, 1969); Reeves, ed., *Prophetic Rome in the High Renaissance Period: Essays* (Oxford: Clarendon Press, 1992); Ronald G. Musto, *Apocalypse in Rome: Cola di Rienzo and the Politics of the New Age* (Berkeley: Uni-

versity of California Press, 2003). My thanks to Victoria Morse for her expert advice on Joachim.

7. Inghirami, *Ethruscarum Antiquitatum Fragmenta*, †2r: "Perlegi, et obstupui; locum signavi."

8. Fabrizio Borrelli, *Le Saline di Volterra nel Granducato di Toscana* (Florence: Leo S. Olschki, 2000); Angelo Marrucci, "Nützliche Metalle: Steinsalz und Silber," in *Otto der Große und Europa: Volterra von Otto I bis zur Stadtrepublik*, ed. Andrea Augenti (Siena: Nuova Immagine, 2001), 66–72, with exhaustive bibliography.

9. Inghirami, *Ethruscarum Antiquitatum Fragmenta*, †2v: "Hoc [scarith] Vulterram defero, et cum marmoreis monumentis in Theatro Vulterrano a Raphaele Maffaeo anno 1494 repertis confero, eiusdem formae esse characteres comperio." For the inscriptions themselves, see Françoise-Hélène Massa-Pairault, "La stele di 'Avile Tite' da Raffaele il Volterrano ai giorni nostri," *Mélanges de l'École Française de Rome: Antiquité* 103 (1991): 499–528.

10. Inghirami, *Ethruscarum Antiquitatum Fragmenta*, ††ii v: "Ego Prosperus a Patre meo Vesulio Auspiciis eruditus qui Ethruscis mos est, ut ex monumentis antiquorum accepi venturum Magnum Regem, a quo anni numerentur, sic ab eo numeravi in Vaticiniis meis annos, in quibus ea implenda cognovi ex auspiciis. At Patre meo praemortuo, ut augurandi Artem optime perciperem, Vulterram petii, quod in ea Civitate sit Auspiciorum Collegium; sed quo tempore Catilina a Patribus defecit, a Vulterrani huius Arcis Custos sum destinatus."

11. A recent description of the Catilinarian conspiracy can be found in Anthony Everitt, *Cicero: The Life and Times of Rome's Greatest Politician* (London: John Murray, 2001; reprint, New York: Random House, 2003), 87–112. The chief ancient sources for this period are Sallust (Gaius Sallustius Crispus), *The Conspiracy of Catiline*, trans. S. A. Handford (Harmondsworth: Penguin Classics, 1963); and Cicero's four *Catilinarian Orations*, translated as *Selected Political Speeches of Cicero*, by Michael Grant (Harmondsworth: Penguin Classics, 1969), among the liveliest works of ancient Roman literature.

12. Cicero, *Catilinam*, IV.24; a complete English translation of the Oration can be found in Grant, *Selected Political Speeches of Cicero.*

13. Inghirami, *Ethruscarum Antiquitatum Fragmenta*, ††ii v: "Cum Romani Exercitus Fesulam, et Vulterram Romano Imperio iterum reddidissent, et Castrum hoc, et Arcem obsidione urgerent, et salutem penitus desperarem, charos, familiaresque, Penates, et quam pecuniam penes me habebam cum ea, quae in publico aerario Arcis, et Castri asservabatur, et fatalia Cervi cornua ex auro conflata, non longe ab his antiquis, et pretiosissimis monumentis, et Vaticiniis Ethruscis, et Latinis Characteribus obsignatis reposui, sed quoniam Ethrusca lingua pene obsolevit; ideo quae Ethruscis caracteribus sunt scripta in compendium redegi. . . . Ego Prosperus Fesulanus Augur Anno post Catilinam."

14. Cornelius Tacitus, *Agricola*, 30: "atque ubi solitudinem faciunt, pacem appellant."

15. D. P. Walker, *The Ancient Theology: Studies in Christian Platonism from the Fifteenth to the Eighteenth Century* (London: Duckworth, 1972). Skepticism about *prisca theologia* had already begun to emerge in the works of antiquarians like Ole Worm, however; see Peter N. Miller, *Peiresc's Europe: Learning and Virtue in the Seventeenth Century* (New Haven: Yale University Press, 2000).

16. Giovanni Cipriani, *Il mito etrusco nel Rinascimento fiorentino* (Florence: Leo S. Olschki, 1980); Gabriele Morolli, *"Vetus Etruria": Il mito degli Etruschi nella letteratura architettonica nell'arte e nella cultura da Vitruvio a Winckelmann* (Florence: Alinea Editrice, 1985).

17. Inghirami, *Ethruscarum Antiquitatum Fragmenta*, ††ii 2r.

18. Ibid.

19. Ibid.: "marmorea frusta, ferrum consumptum rubigine, hominum ossa, et marcida, et combusta."

20. This friend was almost certainly Raffaello Maffei; see chapter III.

21. Inghirami, *Ethruscarum Antiquitatum Fragmenta*, 1: "Pater meus . . . me non solum Ethruscam, sed etiam Graecam, et Hebraicam linguam docuit, postea augurandi artem, et ipsius naturae arcana; quae omnia hominum causa fecit. . . . Non tamen hominum cogunt Auguria; nam Maximus Esar, cum hominem creasset, eum suae

voluntatis liberum possessorem constituit...." Ibid., 3: "nam haec tria astra Caris, Mor, et Turg simul coniuncta vidi; vidi etiam ex Asghariis fulguribus, quod omne Scarith habentur, nisi tu, amicus, et pater adfueritis, non invenientur, sed etiam scito, quodsi tu, amicus, et pater ... tribus vobis pessima astra futura."

22. Massimo Pallottino, *Testimonia Linguae Etruscae*, 2nd ed. (Florence: La Nuova Italia, 1968), #803; cf. Pallottino, *Thesaurus Linguae Etruscae*, vol. I (Rome: C.N.R., Centro di Studio per l'Archeologia Etrusco-Italica, 1968), 415. The word is described in Suetonius, *Life of Augustus*, 97.

23. For a superb view of the Thirty Years' War with particular emphasis on some of the characters in the present story, see Konrad Repgen, *Dreißigjähriger Krieg und Westfälischer Friede: Studien und Quellen* (Paderborn: Ferdinand Schöningh, 1998).

24. Maurice A. Finocchiaro, *The Galileo Affair: A Documentary History* (Berkeley: University of California Press, 1989), 291.

25. Renieri's essay, *Monopanthon, De Ethruscarum Antiquitatum fragmentis Scornelli prope Vulterram repertis Disquisitio Astronomica*, was published in 1638 by Amadore Massi in Florence. The text is also published entire in Niccolò Maria Lisci, *Documenti raccolti dell'illustrissimo Signor Canonico Niccolò Maria Lisci ... intorno alle antichità Toscane di Curzio Inghirami* (Florence: Pietro Gaetano Viviani, 1739), 105–8.

26. A bilingual inscription from Pesaro, now in that city's Museo Oliveriano, *Corpus Inscriptionum Etruscarum* [CIE] 7697, commemorates an Etruscan, Larth Cafates (Lucius Cafatius in Latin), who boasts two professional qualifications in Latin—*haruspex* (entrails-diviner) and *fulguriator* (lightning-diviner)—which seem to match a three-word Etruscan phrase: *netśvis: trutnut: frontac*. *Netśvis* is attested elsewhere (CIE 978; spelled *netsviś*), as is *trutnut* (cf. CIE 5487), and both words presumably mean "diviner" of some sort, but their exact meaning is so far unclear; *frontac* (or *fronta*, if -c is a suffix) is as yet unique, and anomalous for its letter "o."

27. Kurt Johannesson, *The Renaissance of the Goths in Sixteenth-Century Sweden: Johannes and Olaus Magnus as Politicians and Historians*,

trans. James Larson (Berkeley: University of California Press, 1991).

28. Olaus Magnus, *Historia de gentibus septentrionalibus, earumque diversis statibus, conditionibus, moribus, ritibus, superstitionibus, disciplinis . . . Authore Olao Magno Gotho* (Rome: apud Ioannem Mariam de Viottis parmensem, 1555), was already translated into Tuscan vernacular by 1565 (*Historia delle genti et della natura delle cose settentrionali da Olao Magno . . . descritta in xii libri. Nuovamente tradotta in lingua toscana* [Venice: Appresso i Giunti, 1565]); see also Bonaventura Vulcanius, *De literis et lingua Getarum* (Louvain: Plantin, 1597).

29. See chapter IV.

30. Inghirami, *Ethruscarum Antiquitatum Fragmenta,* ††ii 2v: "in eam ivimus sententiam, Romanos, Scornello capto, illum abstulisse, deinde Castrum funditus evertisse."

Chapter Two

1. Inghirami, *Ethruscarum Antiquitatum Fragmenta*, 273: "Vultur, et Coluber simul iungentur: nascentur ex eis Turtures: quae ut Aquilae volabunt, sed non attingent finem: expectabunt rorem, et non erit ros; nec pluviae eas saturabant; verum Tirrenum mare gustabunt."

2. The following picture of Tuscan society, indeed the whole of the present book, owes a profound debt to Eric Cochrane's marvelous *Florence in the Forgotten Centuries, 1527–1800* (Chicago: University of Chicago Press, 1973), for its witty writing as well as its erudition. For the Tuscan academies in particular, see Cochrane, *Tradition and Enlightenment in the Tuscan Academies, 1690–1800* (Rome: Edizioni di Storia e Letteratura, 1961). See also Mario Biagioli, *Galileo Courtier* (Chicago: University of Chicago Press, 1993). For the Lincei, see David Freedberg, *The Eye of the Lynx: Galileo, His Friends, and the Beginnings of Modern Natural History* (Chicago: University of Chicago Press, 2002). The Royal Society, founded in 1660, and now the official academy of sciences for the British Commonwealth, maintains a website with its history at http://www.royalsoc.ac.uk/framer.asp?page=/royalsoc/rshist.htm.

3. In the eighteenth century, the Sepolti began to link their name to the fact that they were enthusiastic archaeological excavators, literally as well as figuratively buried.

4. The single known Etruscan linen book was used to wrap a mummy of a mature Egyptian woman who may have been part of an Etruscan trading colony in the Nile Delta. Her mummy was bought in the mid-nineteenth century by a Croatian official in the Austro-Hungarian government, Mihael Baric, who took it home to Zagreb and bequeathed it to the local museum in his will of 1862. When curators unwrapped the mummy in the twentieth century, its markings could be identified as an Etruscan text that seems to be a ritual calendar for the spring and summer months. Good images of Etruscan linen books from tombs as well as a complete photograph and transcription of the Zagreb mummy wrap are presented in Francesco Roncalli, ed., *Scrivere Etrusco* (Milan: Electa, 1985). A concise account of the mummy wrap and recent bibliography can be found in Giovannangelo Camporeale, *Gli Etruschi, storia e civiltà* (Turin: UTET, 2000), 191, 204–6.

5. Inghirami, *Ethruscarum Antiquitatum Fragmenta*, ††ii 2v: "Pisis allatis, illius almi Gymnasii eruditissimi advocantur, qui eas inspiciant. Ipsorum sententiae diversae fuere; agnoscebant quidam antiquitatem, et fatebantur, quidam dolo malo suppositas suspiciantes asserebant, apud Ethruscos veteres chartam linteam non extitisse."

6. Ibid.: "Cum his omnibus Florentiam petii et Magno Duci, Serenissimis Principibus, universae Aulae, et eruditissimis viris palam facio. Nonnulli antiquitatem admirantes rimabantur [sc. mirabantur] materiam Scarith, que extrinsecus ab aqua, intrinsecus ab igne defendebat; at nonnulli, prout hominum varia sunt ingenia, fictitias, et suppositas dicebant." The "bronze lamp" is the bow of a genuine Etruscan fibula; see chapter IX. The household god, with its false Etruscan inscription, is surely a forgery; furthermore its composition is reported as tin in the *Ethruscarum Antiquitatum Fragmenta* and as lead elsewhere.

7. Enrico Fiumi, *L'impresa di Lorenzo de' Medici contro Volterra (1472)*

(Florence: Leo S. Olschki, 1948). The leader of the Florentine troops, Federico da Montefeltro, Count of Urbino, became duke for this cruel victory.

8. See Furio Diaz, *Il Granducato di Toscana—I Medici* (Turin: UTET, 1987), with extensive bibliography.

9. See Mario Battistini, *L'ammiraglio Jacopo Inghirami e le imprese dei cavalieri dell'Ordine di S. Stefano contro i Turchi nel 1600* (Volterra: Tip. Confortini, 1912).

10. Diaz, *Il Granducato*, 292–94, 301–3.

11. Umberto Bavoni, *La Cattedrale di Santa Maria Assunta e il Museo Diocesano di Arte Sacra di Volterra* (Florence: Edizioni IFI, 1997), 55–62.

12. Mons. Maurizio Cavallini, *Guida: Cattedrale Volterra. Battistero* (Volterra: UTA, 1957).

13. By 1637 she was engaged to Bernardo Picchinesi, as Curzio notes in the Inghirami family tree appended to *Ethruscarum Antiquitatum Fragmenta* (not paginated, it, like all the other added plates, appears in different places in different copies of the book).

14. Cavaliere Francesco also held public office, as Prior of Borgo San Sepolcro.

15. Diaz, *Il Granducato*, 85–229; Cochrane, *Florence in the Forgotten Centuries*, 13–92.

16. Roberto Ferretti, "La pirateria barbaresca sulle coste della Maremma," in *I Medici e lo stato senese, 1555–1609: Storia e territorio*, ed. Leonardo Rombai (Rome: De Luca, 1989), 40–41; Viviano Domenici, "La ragazza rapita che conquistò il sultano," *Corriere della Sera* 3 (August 2003).

17. See Andrea Galdy, "'Con bellissimo ordine': Antiquities in the Collection of Cosimo I de' Medici and Renaissance Archaeology" (Ph.D. diss., University of Manchester, 2002).

18. Diaz, *Il Granducato*, 423–63.

19. Ibid., 367–464.

20. Cavaliere Giulio Inghirami, letter to Lord Bailiff Andrea Cioli, 29 March 1635, BAV, MS Barb. Lat. 3150, c. 264r–v: "Questa sera li Signori Inghiramo Inghirami, et Curzio Inghirami questo mio Cug-

ino, et questi mio Nipote sono andati a Palazzo per mostrare al Serenissimo Priore le rare memorie trovate da Loro in una Villa, che mio Padre cambiò con essi loro. . . . Vi sono più carte scritte di Carattere Toscano, o per dir meglio Etrusco antico, et se bene fra questi eruditi è stata contesa, se la Carta nostrale fosse in quei tempi, l'esperienza insegna, che la Carta Lintea [264v] accennata da Livio, e da Plinio, è quella che usiamo hoggi noi."

21. Ibid.: "le quali [memorie] cominciano da Noè fondator di Volterra, et contengono per 1800 anni la serie continuata di 55 Re Toscani, la fondazione delle 12 Città, della Toscana. . . . [264v] Vi sono predizioni ancora della venuta, vita, et morte di Nostro Signore Gesù Christo più chiare che non si lassano nella Biblia, et fra l'altre, che perceperat: Venturum Magnum Regem a quo anni numerentur."

22. Walter E. Stephens, "Berosus Chaldaeus: Counterfeit and Fictive Editors of the Early Sixteenth Century" (Ph.D. diss., Cornell University, 1979); Stephens, *Giants in Those Days: Folklore, Ancient History, and Nationalism* (Lincoln: University of Nebraska Press, 1989); Anthony Grafton, *Forgers and Critics: Creativity and Duplicity in Western Scholarship* (Princeton: Princeton University Press, 1990); Marianne Wifstrand Schiebe, "Annius von Viterbo und die schwedische Historiographie des 16. und 17. Jahrhunderts," *Skrifter Utgivna av Kungliga Humanistiska Vetenskaps-Samfundet i Uppsala* (*Acta Societatis Letterarum Humaniorum Regiae Upsaliensis*) 48 (1992): 7–26; Edoardo Fumagalli, "Un falso tardo-quattrocentesco: Lo Pseudo-Catone di Annio da Viterbo," in *Vestigia, Studi in onore di Giuseppe Billanovich*, ed. Rino Avesani (Rome: Edizioni di Storia e Letteratura, 1984), 337–83; Fumagalli, "Aneddoti della vita di Annio da Viterbo, O.P., 1. Annio la vittoria dei genovesi sui sforzeschi; 2. Annio e la disputa sull'Immacolata Concezione," *Archivum Fratrum Predicatorum* 50 (1980): 167–99; O. E. Danielsson, "Annius von Viterbo über di Gründungsgeschichte Roms," in *Corolla archaeologica prinicipi hereditario regni sueciae Gustavo Adolpho dedicata* (Lund: Glerup, 1932), 1–16; Ingrid Rowland, "*L'Historia Porsennae* e la conoscenza degli Etruschi nel rinascimento," *Res Publica Litterarum* 9 (1989): 185–93.

23. For further discussion of Annius of Viterbo, see chapter VIII.

24. Inghirami to Cioli, 29 March 1635, MS Barb. Lat. 3150, c. 264v: "Credo che il Proveditore Serenissimo ce le lasserà stampare poiché sono cose trovate nella nostra Villa posseduta dalla Casa Nostra più di 450 anni continui, senza che mai da nostri Vecchi si sia saputo che vi fosse anticamente, o Villa, o Rocca, né tampoco si sarebbe saputo, che il caso non havesse portato, che il Signor Curzio suddetto et la Signorina Lucrezia sua sorella, trovassero un involto di pece, pelo, et calcestruzzo, dentro al quale era una Carta che diceva Thesaurum invenisti, signa locum, et abi."

25. Ibid., 31 March 1635, c. 264v–265r: "Stasera è stata l'ultima sessione sopra le scritture di Volterra, et finalmente fra questi Signori Critici si conclude, che sieno le più nobili Scritture di Manoscritti, che si trovino hoggi in Europa a nostra notizia. Il Serenissimo Provveditore si contenta che si stampino, et noi faremo in Volterra fare il Processo del modo, del tempo, et da chi furono ritrovate et riconoscerle, et questo medesimo faremo poi far qui in Fiorenza, et alla Ruota, et da Monsignore Nunzio, et in Lamine di Rame, secondo che sono grandi i pezzi di carta dove sono scritte, s'imiteranno quei Caratteri il più che si possa [265r] affinché il Mondo goda questo sí nobil tesoro che doverà esser grato all'Italia poiché vi è la sua origine dal Diluvio in qua."

26. Gerald Tyson and Sylvia Wagonheim, eds., *Print and Culture in the Renaissance: Essays on the Advent of Printing in Europe* (Newark: University of Delaware Press, 1986); M. B. Parkes, *Pause and Effect: An Introduction to the History of Punctuation in the West* (Aldershot: Scolar Press, 1992); Alberto Asor Rosa, ed., *Letteratura italiana*. Vol. 2: *Produzione e consumo* (Torino: Einaudi, 1983), 499–524; Nick Wilding, "Writing the Book of Nature: Natural Philosophy and Communication in Early Modern Europe" (Ph.D. diss., European University Institute, 2000). Thanks also to Thomas Willette and Rodney Palmer for their expertise on this subject.

27. Capponi's interest in Scornello may have extended beyond Etruscan antiquities, given the villa's, and Inghiramo Inghirami's, involvement with the extraction of salt in Volterra.

28. See the trial records compiled in Volterra, Biblioteca Comunale Guarnacci, MS LII.6.5, partially transcribed in Lisci, *Documenti raccolti*, 36–77.

29. Curzio Inghirami, *Discorso sopra l'opposizioni fatte all'antichità toscane* (Florence: Amadore Massi and Lorenzo Landi, 1645), 15: ". . . la medesima Altezza, che mandò a posta a riconoscere il luogo, e vedere cavare, i Signori Mario Guiducci, e Niccolò Arrigheti, Gentiluomini Fiorentini, il qual protesto si conserva nell'Archivio della Città di Volterra." The document to which he refers is now preserved in Volterra, Biblioteca Comunale Guarnacci, MS LII.6.5.

30. The original MS, covered with printer's ink, is in Volterra, Biblioteca Comunale Guarnacci, MS LII.6.1.

31. See the flyleaf of the original MS: "typis edita Florentiae anno 1636 fol. rincusa Francofurti fol. anno 1637." As Thomas Cerbu has suggested in a private communication, Massi's initials may appear upside down in the Etruscan letters on the title page.

32. His first wife, Giulia Incontri, died in 1637; he married his cousin Maria Inghirami in September 1638, but he died in May 1639. Giovanni Batistini, "Il ritratto di Fedra Inghirami, opera di Raffaello: Il vero e il falso," *Rassegna Volterrana* 71–72 (1994–95): 68–69, 75.

33. See chapter VII.

Chapter Three

1. See John Beldon Scott, *Images of Nepotism: The Painted Ceilings of Palazzo Barberini* (Princeton: Princeton University Press, 1991).

2. For some vivid accounts of Barberini Rome, see Pietro Redondi, *Galileo Eretico* (Turin: Einaudi, 1983); Torgil Magnuson, *Rome in the Age of Bernini* (Stockholm: Almqvist and Wiksell, 1982 [vol. I]; 1986 [vol. II]); Scott, *Images of Nepotism*; Patricia Waddy, *Seventeenth-Century Roman Palaces: Use and the Art of the Plan* (New York: Architectural History Foundation/Cambridge: MIT Press, 1990); Sarah McPhee, *Bernini and the Bell Towers: Architecture and Politics at the Vatican* (New Haven: Yale University Press, 2002); Jennifer Montagu, *Roman Baroque Sculpture: The Industry of Art* (New Haven:

Yale University Press, 1998); Francis Haskell, *Patrons and Painters: A Study in the Relations between Italian Art and Society in the Age of the Baroque*, rev. 2nd ed. (New Haven: Yale University Press, 1980); Paolo Portoghesi, *Francesco Borromini: Architettura come linguaggio* (Milan: Electa, 1967); and Frederick Hammond, *Music and Spectacle in Baroque Rome: Barberini Patronage under Urban VIII* (New Haven: Yale University Press, 1994). See also Cochrane, *Florence in the Forgotten Centuries*, 165–228.

3. John W. O'Malley, *The First Jesuits* (Cambridge: Harvard University Press, 1993). The vow of obedience specifically regards willingness to go on a mission at the pope's command.

4. See Eugenio Lo Sardo, ed., *Athanasius Kircher, S.J., Il Museo nel Mondo* (Rome: De Luca, 2001); Lo Sardo, *Iconismi et Mirabilia da Athanasius Kircher* (Rome: Edizioni dell'Elefante, 1999); Daniel Stoltzenberg, ed., *The Great Art of Knowing: The Baroque Encyclopedia of Athanasius Kircher* (Stanford: Stanford University Libraries/Fiesole: Cadmo, 2001); Ingrid D. Rowland, *The Ecstatic Journey: Athanasius Kircher in Baroque Rome*, with an introduction by F. Sherwood Rowland (Chicago: Dept. of Special Collections, University of Chicago Library, 2000); Dino Pastine, *La nascita dell'idolatria: l'Oriente religioso di Athanasius Kircher* (Florence: La Nuova Italia, 1978); Thomas Leinkauf, *Mundus combinatus: Studien zur Struktur der barocken Universalwissenschaft am Beispiel Athanasius Kirchers SJ (1602–1680)* (Berlin: Akademie-Verlag, 1993); Maristella Casciato, Maria Grazia Ianniello, and Maria Vitale, eds., *Enciclopedismo in Roma barocca: Athanasius Kircher e il museo del Collegio Romano tra Wunderkammer e museo scientifico* (Venice: Marsilio, 1986); and Wilding, "Writing the Book of Nature."

5. For Sacchi's *Allegory of Divine Wisdom*, see Scott, *Images of Nepotism*, 38–67, with bibliography.

6. Galileo Galilei, *Dialogo sopra i due massimi sistemi del mondo tolomaico e copernicano*, ed. Libero Sosio (Turin: Giulio Einaudi editore, 1970), 548: "So che amendue voi, interrogati se Iddio con la Sua infinita potenza e sapienza poteva conferire all'elemento dell'acqua il reciproco movimento che in esso scorgiamo, in altro modo

che co'l far muovere il vaso contenente, so, dico, che risponderete, avere egli potuto e saputo ciò fare in molti modi, ed anco dall'intelletto nostro inescogitabili. Onde io immediatamente vi concludo che, stante questo, soverchia arditezza sarebbe se altri volesse limitare e coartare la divina potenza e sapienza ad una sua fantasia particolare."

7. See editor's note by Libero Sosio, Galileo, *Dialogo supra i due massimi sistemi*, 548–50.

8. Galileo, *Dialogo sopra i due massimi sistemi*, 549: "Mirabile e veramente angelica dottrina." The biliography on Galileo's trial is vast, including Giorgio di Santillana, *The Crime of Galileo* (Chicago: University of Chicago Press, 1955); Annibale Fantoli, *Galileo, for Copernicanism and for the Church* (translation by George V. Coyne of *Galileo, per il copernicanesimo e per la chiesa*, 2nd ed., rev. and corr. [Vatican City]: Vatican Observatory Publications, 1996); Richard Blackwell, *Galileo, Bellarmine, and the Bible* (Notre Dame: University of Notre Dame, 1992); Stillman Drake, *Essays on Galileo and the History and Philosophy of Science* (Toronto: University of Toronto, 1999); Stillman Drake, *Galileo at Work: His Scientific Biography* (Chicago: University of Chicago Press, 1978); Maurice Finocchiaro, trans. and ed., *The Galileo Affair: A Documentary History* (Berkeley: University of California Press, 1989); Jerome Langford and Stillman Drake, *Galileo, Science and the Church* (New York: Desclee, 1966); Lawrence S. Lerner and Edward A. Gosselin, "Galileo and the Specter of Bruno," *Scientific American* 255, no. 5 (November 1986): 126–33; Peter Machamer, ed., *The Cambridge Companion to Galileo* (Cambridge: Cambridge University Press, 1998); Michael John Gorman, "A Matter of Faith?: Christoph Scheiner, Jesuit Censorship, and the Trial of Galileo," *Perspectives on Science* 4 (1996): 283–320; Maurice A. Finocchiaro, "Philosophy versus Religion and Science versus Religion: The Trials of Bruno and Galileo," in *Giordano Bruno, Philosopher of the Renaissance*, ed. Hilary Gatti (Aldershot: Ashgate, 2002), 51–96; and Tommaso Campanella, *Apologia pro Galileo*, ed. and trans. Jean-Michel Lerner (Paris: Les Belles Lettres, 2001).

9. Acts and Decrees of the Fourth Session of the Council of Trent, *Concilii Tridentini Actarum Vol. V, Pars Altera: Acta post sessionem tertiam usque ad Concilium Bononiam translatum*, ed. Stephan Ehses (Freiburg i. Breisgau: B. Herder, 1911), 92: "Praeterea ad coercenda petulantia ingenia decernit [sc. Concilium] ut nemo, suae prudentiae innixus, in rebus fidei et morum ad aedificationem doctrinae Christiane pertinentium sacram scripturam ad suas contorquens, contra eum sensum quem tenuit et tenet sancta mater ecclesia . . . ipsarum scripturarum sacrarum interpretari audeat." See also John W. O'Malley, *Trent and All That: Renaming Catholicism in the Early Modern Era* (Cambridge, MA: Harvard University Press, 2000).

10. Galileo Galilei, *Le Opere di Galileo Galilei*, Edizione Nazionale, ed. Antonio Favaro (Florence: Barbera, 1968), vol. XIX, 406–7.

    . . . *Io Galileo, figliuolo del quondam Vincenzo Galileo di Fiorenza, dell'età mia d'anni 70, costituto personalmente in giudizio, e inginocchiato avanti di voi Eminentissimi e Reverentissimi Cardinali, in tutta la Republica Cristiana contro l'eretica pravità generali Inquisitori; avendo davanti gl'occhi miei li sacrosanti Vangeli, quali tocco con le proprie mani, giuro che sempre ho creduto, credo adesso, e con l'aiuto di Dio crederò per l'avvenire, tutto quello che tiene, predica e insegna la Santa Cattolica e Apostolica Chiesa.*

    Ma perché . . . dopo d'essermi stato con precetto dall'istesso [Santo Offizio] giuridicamente intimato che omninamente dovessi lasciar la falsa opinione che il sole sia centro del mondo e che non si muova e che la terra non sia centro del mondo e che si muova, e che non potessi tenere, difendere né insegnare in qualsivoglia modo, né in voce né in scritto, la detta falsa dottrina, e dopo d'essermi notificato che detta dottrina è contraria alla Sacra Scrittura, scritto e dato alle stampe un libro nel quale tratto l'istessa dottrina già dannata e apporto ragioni . . . a favor di essa . . . sono stato giudicato veementemente sospetto d'eresia . . . con cuor sincero e fede non finta abiuro, maledico e detesto li sudetti errori e eresie.

11. Ibid., 402–7.

12. Vincenzo Noghera, letter to Cassiano dal Pozzo, 18 September 1638, Rome, Biblioteca Corsiniana, Archivio dal Pozzo, MS. XII (10), c. 458r; cf. c. 430v, 9 September 1637. For Noghera's experi-

ence with old documents, see BAV, MS Chigi G.II.65, c. 232v. For his Portuguese connections, see A. J. Lopes de Silva, *Cartas de D. Vicente Nogueira, publicadas pelo director da Biblioteca pública de Évora A. J. Lopes da Silva* (Coimbra: Impr. da universidade, 1925).

13. Ingrid D. Rowland, "Etruscan Inscriptions from a 1637 Autograph of Fabio Chigi," *American Journal of Archaeology* 93 (1989): 423–28. The article requires correction in the final note: in 1989 Thomas Cerbu and I erroneously believed, following one of the early librarians of the Fondo Chigi, that Noghera was the inventor of the pseudonymous author Benno Durkhundurk; Noghera is still listed as Benno's inventor in the card catalog and online OPAC of the Vatican Library. Thomas Cerbu's discovery of the real identity of that personage is discussed below, in chapter V.

14. See chapter V.

15. Vincenzo Noghera, letter to Cardinal Sacchetti, copy for Fabio Chigi, BAV, MS Chigi G.II.65, 231v: "Se dunque provarò che queste Antichità no' forno scritti nel tempo che dicono elleno stesse, ne fu possibile, che all hora si scrivessero nè anco più di cento e venti anni doppo, ben si consegue, che ne anche furono scritte, e raccolte dall'Autor che dicono: e consequentemente, che è un impostura, et una fintione, che niente ha di buono." Cardinal Sacchetti is identified as the addressee on c. 238v.

16. Freedberg, *The Eye of the Lynx*; Francesco Solinas, ed., *Cassiano dal Pozzo: Atti del Seminario Internazionale di Studi* (Rome: De Luca, 1989); Ingo Herklotz, *Cassiano dal Pozzo und die Archäologie des 17. Jahrhunderts* (Munich: Hirmer, 1999).

17. The Barberini circle also maintained contact with the Swiss scholar Paganino Gaudenzio, who was a professor at Pisa, and the Danish jurist Heinrich Ernst, both of whom would eventually publish their opinions of the scarith; see chapter IV. A letter from Ernst to Lucas Holstenius (at that time Cardinal Francesco Barberini's librarian) from November 1637 mentions a recent trip to Rome, BAV, MS Barb. Lat. 2180, c. 235: "cum Romae familiariter sum usus." Paganino Gaudenzio is in close contact with Fabio Chigi, see BAV, MS Chigi R.I.18.

18. This emerges in Noghera's letter to Sacchetti, 235r.

19. Guillaume Postel, *De Etruriae regionis quae prima in orbe Europaeo habitata est, Originibus, Institutis, Religione et Moribus . . . et imprimis De Aurei Saeculi Doctrina commentaria* (Florence, 1551); edited by Giovanni Cipriani, *Guillaume Postel, De Etruriae regionis quae prima in orbe Europaeo habitata est, Originibus, Institutis, Religione et Moribus* (Florence: Consiglio Nazionale delle Ricerche, 1986).

20. Noghera to Sacchetti, 231r–237v; see 235r: "Pur io mai crederò che il Postello fusse l'Autore, non perche dalla sua conscienza non potesse sospettarsi ogni tristitia e furfanteria, essendo processato di Magia e Stregheria, e non so se coniunto, et anche di cativa vita e costumi, ma perche dalla sua scienza e sapere ho gran Notitia, per haver lecto parecchi libri suoi, quantunque prohibiti . . . in tuti i quali scopre il Postello una dottrina molto soda, et esser nelle Lingue orientali un prodigio . . ."; 235v: "Postello . . . mentirebbe come discreto, non come sciocco, non trovandosi foglio, ne ancho pagina, contra la quale non possino trovarsi inconvenienti."

21. Ibid., 234r: "Et essendo tanto il numero de gli Historici, et altri Interlocutori, che qui s'introducono . . . non si scorge nel loro stile una minima differenza, anzi tanto uniformità e consonantia, come si fiorissero tutti nell'istesso tempo, o per dir meglio esser tutti scritti da una mano, e d'una istessa penna, et è tutta opera giovenile."

22. Inghirami, *Ethruscarum Antiquitatum Fragmenta*, †ii 3v.

23. Noghera to Sacchetti, 234v: "Vedasi se poteva più patentemente disegnarsi il Gentilhomo che trovo questo Tesoro quantunque lo nominassi, che con tanti segni individuali e' haver Padre Signore del Castello e virtuoso, esserlo egli, et il amico suo. . . ."

24. Ibid.: "omnem sermonem qui dilucide explicat personam de qua quis loquitur haberi pro nominatione."

25. Noghera to dal Pozzo, c. 452v: "il grado che havea dal suo Padre, et il mio dall'avo. . . ."

26. Vincenzo Noghera, BAV, MS Barb. Lat. 6472. The Vatican Library's catalog for this manuscript says:

Francesco di Noya = V. Noghera.

He was a Portuguese nobleman who had lost his reputation in his own country and came to Rome. He went into service with Cardinal Sacchetti, Legate in Bologna. When he returned from Bologna to Rome he aspired to become part of Cardinal Francesco's court, and also to accompany the Cardinal in France as an advisor, for, as he says in his letter of 11 March 1646, I regard myself as inferior to no one. *He never obtained the desired positions, although he had monetary help and a place to live in Palazzo della Cancelleria. On 4 February 1646 he says he has entered his sixtieth year. He was most erudite, but he prided himself too much on his knowledge and was too free in his habits; nor could it have pleased Barberini in the midst of all his concerns to have had a Portuguese Spaniard in his household, a name that Noghera applies to himself in the letter of 23 September 1641. Noghera, in my judgment, was a little version of Tommaso Campanella. . . . In his last letter, it is evident that he was threatened with imprisonment.*

The threat of prison with which the Barberini correspondence ends so dramatically in 1648 turned out to be a false alarm. Noghera survived for another four years as an active informant for King João IV of Portugal.

[Era nobile Portoghese che aveva perduto la buona fama in patria ed era venuto a Roma. Andò al servizio del Card. Sacchetti Legato a Bologna. Tornato da Bologna a Roma ambiva esser uno della corte del Card. Francesco, ed anche di trovarsi col medesimo in Francia in qualità di consigliere, poiché come dice nella lettera dell' 11 marzo 1646, non mi tengo inferior a nessuno. *Non ebbe mai il posto desiderato, ebbe però aiuti pecuniari e l'abitazione nel Palazzo della Cancelleria. Il 4 febbraio 1646 dice di essere entrato nell'anno 60. Era molto erudito, ma troppo pretendeva a sapere ed era troppo libero, né al Barberino nelle sue vicende doveva aver piacere di avere presso di sé uno Spagnolo Portoghese, nome che il Noghera dà a sé stesso nella lettera 23 Settembre 1641. Il Noghera, a mio giudizio, era una piccola immagine di Tommaso Campanella. . . . Apparisce dell'ultimo biglietto, che gli era minacciata la prigione.]*

27. See note 26.

28. BAV, MS Barb. Lat. 6472.

29. Noghera to dal Pozzo, c. 452v: "De' Signori Inghirami son servitor, e assai amico del Cavalier Giulio, e di tutto il loro trattato, ma ve mi dispiacque tanto, come voler meter in scena senza besogno e anche senza luogo la memoria di quel Heroe. Se non fussero gentilomini e persone nobili, haverei pensato che l'havessero fatto con mala Fede ma essendo quelli che sono voglio pensar che veramente intendessero esser offesa quella Altezza."

30. Noghera to dal Pozzo, Bologna, 14 August 1638, c. 456r.

31. Ibid.: "le parole picanti, sempre le intesi impersonalmente et in rem, senza pensar né pecato veniale nel Signor Curtio (credulità nimia sí) altrimenti sarebbono restate nel calamaio perché ho io con questa nobilissima familia un grandissimo rispetto."

32. Ibid., Bologna, 18 September 1638, c. 458r: "Mille volte mi son pentito di haver pigliato la penna sopra le antichità di Volterra bastando per sodisfar l'obedienza dil Signor Cardinal mio dirgli di parola il mio senso pur essendosi piacciuto la scrittura tale quale e domandandomi consenso da farlo redder a' Signori Inghirami, facilmente il prestai, perché come mai sospetai impostura in questi Signori, cosí non pensava che eglino havessero di purgarsi d'essa ma subito che notesi il contrario, raccolsi dal Signor Cardinale non solo la scrittura originale ma anche la risposta e tutte due ho tenuto e tengo serrate sotto c[h]iave senza che usasse copia nessuna."

33. Batistini, "Il ritratto di Fedra Inghirami," 69, 74.

Chapter Four

1. Leone Allacci, *Leonis Allatii Animadversiones in Antiquitatum etruscarum fragmenta ab Inghiramio edita* (Paris: S. Cramoisy, 1640), 31.

2. See Giuseppe Godenzi, *Paganino Gaudenzi* (Bern: H. Lang/Frankfurt: P. Lang, 1975); Giampiero Brunelli, *Dizionario Biografico degli Italiani*, vol. 52 (Rome: Enciclopedia Italiana, 1999), 676–78, s.v. *Gaudenzi, Paganino*, with updated bibliography.

3. The censors have noted their approval on the back of the autograph manuscript of Gaudenzio's *De Charta: Exercitatio*, BAV, MS Urb. Lat. 1605, 103v, on 25 August 1636, 9 September 1636, and 15 September 1636.

4. Paganino Gaudenzio, *De Charta*, BAV, MS Urb. Lat. 1605, 74r: "Mirum profecto est reperiri hac tempestate homines, qui affirmare ausint Romanis non ignotam fuisse tempore Syllae, chartam qualem non habemus ex contritis laceratisque linteolis factam."

5. Ibid., 79r: "clamabo enim et riclamabo [*sic*] antiquam chartam ex papyro confici solitam."

6. Ibid., 83r: "Aptissime autem vocavimus chartam, quia magnum habet charta nostra cum charta, quae ex papyro conficiebatur, affinitatem."

7. Paganino Gaudenzio, *Paganini Gaudentii in Antiquitates quasdam editas sub nomine Prosperi Faesulani Animadversio*, BAV, MS Urb. Lat. 1605, c. 115v: "Quis figmentum Prosperi non agnoscat? Quis non explodat audaciam?"; c. 130r: "Ohé iam satis est, ohé lector." Guadenzio is here quoting the opening line of Martial's epigram IV.89: "Ohé, iam satis est, ohé, libellus" ("Oy, little book, oy, enough already").

8. See Ingrid D. Rowland, *The Culture of the High Renaissance: Ancients and Moderns in Sixteenth-Century Rome* (Cambridge: Cambridge University Press, 1998), 10–22.

9. Contemporaries called the play *Hippolytus*, after its other main character, the chaste stepson who refuses Phaedra's love; ibid., 21, 258–59n27.

10. The anecdote about the wayward stage set is given by a controversial source: none other than Inghirami, *Discorso*, 50. However, every other indication from Tommaso Fedra's life and surviving writings make it seem eminently likely; see Rowland, *High Renaissance*, 21–22, 151–57.

11. Tommaso tried, with only partial success, to change the nickname to "Fedro" or "Phaedrus" in his own day; see Rowland, *High Renaissance*, 21–22. The Inghirami family, however, has unanimously chosen "Fedra" for its boys, usually although not always combined

with Tommaso; there have also been generations of girls named Fedra.

12. Gaudenzio, *Paganini Gaudentii Animadversio*, 122v: "Valde credibile est à Thoma Faedro, qui videtur haec scripta architectatus, memorialem chartam traditam fuisse Inghiramiis, quae doceret unde prima esset petenda charta, ut deinceps ad alia effodienda tenderetur."

13. Ibid., 118v–19r: "Forte etiam suspicatus quis fuerit huic scarith ab ipso defensore additam particulam, quae agit de charta, cum repertum fuerit postquam ego saepius dixissim, nostratem chartam Romanis ignotam fuisse."

14. Ibid., 116v: "Ei rei si inveniat Defensor hominem qui fidem habeat, aut saltem non suspicetur de dolo, tunc ego incredibilium narrationum ingentem numerum exosculabor promptissime."

15. See Cipriani, *Il mito etrusco*.

16. Massimo Pallottino, "Héritages lexicaux," in *Les Étrusques et l'Europe* (Milan: Fabbri, 1992), 246–47.

17. Diaz, *Il Granducato*, 406–7; Cochrane, *Florence in the Forgotten Centuries*, 196–98; Carlo M. Cipolla, *Cristofano and the Plague: A Study in the History of Public Health in the Age of Galileo* (London: Collins, 1973).

18. See Cochrane, *Florence in the Forgotten Centuries*, 165–230, esp. 195–200; Cipolla, *Cristofano and the Plague*.

19. Paganino Gaudenzio, *Epithalamium* for Grand Duke Ferdinando and Vittoria della Rovere, BAV, MS Urb. Lat. 1580, 2r.

20. Heinrich Ernst and Paganino Gaudenzio, *Ad antiquitates etruscas quas Volaterrae nuper dederunt observationes, in quibus disquisitionis astronomicae de etruscarum antiquitatum fragmentis auctor quoque notatur* (Amsterdam: apud Joannem Janssonium, 1639). On the relationship between Ernst and Gaudenzio, see Luc Deitz, "Die Scarith von Scornello: Fälschung und Methode in Curzio Inghiramis 'Ethruscarum antiquitatum fragmenta' (1637)," *Neulateinisches Jahrbuch* 5 (2003): 103–34.

21. Ernst and Gaudenzio, *Ad antiquitates etruscas*, A 2r: "Haut credibile est, quanto cum gaudio librum istum aperui, inspexi, legi; quan-

taque cum indignatione clausi, abieci, devovi. Statim enim deprehendi, fragmenta ista temporibus Syllae, ut videri volunt, conscripta non esse, sed supposititia et ad simplices mortales ludificandos ficta. Idque res ipsa loquitur." The spelling of "haut" is also a peculiarity of the Danish coauthor; Gaudenzio writes *haud.*

22. Ibid., 66: "Quis enim, nisi cui cerebrum in calce est, non videt, omnia ista quae coacervavit, ridicula esse, ficta, contorta. Pag. 4 ex Fragmentis tria astra citat: Caris, Mor, et Turg. Quaeso quaenam hae stellae?"; 69: "recentiores vero, auctore magno nostro Tychone Braheo, uno solo vinculo . . . constellationem hanc pingunt et describunt."

23. Ernst is almost certainly referring here to Fra Mario Giovanelli, O.E.S.A. [Ordo Eremitarum Sancti Augustini], *Cronistoria dell'antichità e nobiltà di Volterra, cominciando dal principio della sua edificazione infin'al giorno d'hoggi* (Pisa: appresso Giovanni Fontani, 1613), whose work Curzio Inghirami disparages in no uncertain terms; see chapter VII.

24. Heinrich Ernst, preface to *Ad antiquitates etruscas*, A 2v–A 3r: "Credo tamen ornatissimum virum Curtium Inghiramium cuncta ista, quae edidit, ita vera esse, sicuti scripta invenerit, existimare; sed longe graditur a sententia mea. Nimis siquidem notum est, multa eiusdem generis alia a civibus Volaterranis excogitata esse, ne fides antiquarii vacillaret. Quibus quare Inghiramius, saepe de ea re monitus, eruditionem suam prostituere voluerit, mirari haut satis queo. Non erat, mea quidem opinione, prudentia boni viri, candidissimam suam simplicitatem sontibus illis, ac fallacibus chartis polliceri."

25. Vincenzo Noghera, letter to Cassiano dal Pozzo, 14 August 1638, Rome, Biblioteca Corsiniana, Archivio dal Pozzo, ms. XII (10), c. 456r: "io hebbi gusto della risposta del Signor Curtio, perché no mi lo levò né accecò il giudicio qualche acerbità sua anzi viddi assai doctrina e molta eruditione, et assai ben sodisfatto il dubio della carta in che hoggi scriviamo pensavo scriverli alcune avvertenze di momento in punti che restono ancora debole, ma . . . [here Noghera characteristically describes all the details of his illnesses]."

26. For a good sketch of Buonmattei's eventful life, see I. Calabresi, *Dizionario Biografico degli Italiani*, vol. 15 (Rome: Enciclopedia Italiana, 1972), 264–68, s.v. *Buonmattei, Benedetto.*

27. Ibid. (see n. 25 supra); Cochrane, *Florence in the Forgotten Centuries*, 203. Romeo the cat, "amatissimo gatto," master of the liberal arts, is praised under Buonmattei's pseudonym, Benduccio Riboboli da Matelica; see Buonmattei, Collected Papers, Florence, Biblioteca Nazionale Centrale, MS Magliabechi IX.122, 318ff.

28. Benedetto Buonmattei to Curzio Inghirami, ibid., c. 167r:

> *Signor mio. Sarei di parere che non toccando l'oppositioni cos'alcuna partiale, o non si dovesse rispondere: o starsi attorno su' generali, e serbar'a miglior occasione le ragioni, che da lei si adducono: perché ella non è l'autor di quelle scritture: e avendole date fuori tali, quali l'ha trovate; non par che sia obbligata a rispondere ad ogni quesito; che possa depender da ogni privato capriccio. Ma è già in possesso: e chi ne la vuole spogliare non bisogna che si fondi su presupposti: ma venga con ragioni chiare, e provanti: e allora si potrà rispondere a suo bell'agio.*
>
> *Ma se pure ella volesse dir qualcos'amichevolmente a chi amichevolmente le ha mandate l'opposizione la consiglierei ad esser più breve: e dove l'oppositor non prova: non risponderei altro che Io non la credo così. A me par tutto il contrario; o cosa tale. E dove le ragioni par che più tosto faccian per lei: le concederei tutte: né soggiugnerei una parola.*

29. Noghera to dal Pozzo, Bologna, 18 September 1638, c. 458r: "... et essendo importunato dal Signor Leone Allatio, et altronde a mandarla risposta a tutti, che io stava tanto sodisfatto dalla dottissima risposta che non mi restava quasi difficultà per sciogler di tutto questo come di verità infallibile può Vostra Signoria accettar il Signor Balì Cioli e che quantumque i Signori Inghirami pubblichino questa vorrei io non lo farò mai né in una minima parola per ciò che desio conservar la gratia et amicitia di quei Signori e perché così conviene allo stato, età, fortuna, e reposo mio."

30. Leone Allacci, *Apes urbanae, sive de viris illustribus, qui ab anno MDCXXX per totum MDCXXXII Romae adfuerunt ac typis aliquid evulgarunt* (Rome: Ludovico Grignani, 1633); Allacci himself

appears under the entry *Leo Allatius*. The book, true to its title, lists the publications of each author.

31. Ibid., p. 5 (A 3r).

32. Leone Allacci, *Socratis, Antisthenis, et aliorum socraticorum epistolae. Leo Allatius hactenus non editas primus graece vulgavit; latine vertit; notas adiecit; dialogum, De scriptis Socratis, praefixit* (Paris: S. Cramoisy, 1637).

33. Allacci, *Animadversiones*, 1: "Vix me, et non sine stomacho a logis sive ridiculis, ac meris nugis expedientem, novum Augiae stabulum, maleolentiae, ac flagitii plenum, retardat. Repentino enim nesciounde irrepentia Etruscarum Antiquitatum Fragmenta, amica titulo, argumento concupita, novitate rara, et suo decore, ornatuque desiderabilia, vix satis otium studio meo suppeditarunt. Sed heu miseram rerum humanarum conditionem! Annon modo magis, quam usquam aliàs appositè dixerim, *magnus liber, magnum malum?* Venenum sub melle est, tribuli, et rubi ubique. . . . An non ego stultus, et improbus, et in me ipsum, eo etiam, qui ludificat, inclementior, si toties ab eo in ipsa Operis fronte monitus, ut caveam, Cave, Cave, Cave, mihi non cavero, sed rebus propositis adeo inauspicato consensum praebuero?"

34. Ibid., 25: "tantum abesse dignoscitur ab aureo illo saeculo, quo Prosper iste scripsisse fingitur, ut eius vel minimum quamque nostri seculi, Historicum, pudere possit; adeo ineptè pleraque, ac ruditer habita, verba, locutio, periodus, structura, dicendi formae, sententiae, ac stylus ipse totus."

35. Ibid., 22: "Uno et eodem tempore legentibus miserandus, et ridendus [Lusor], lacrymas, et cachinnos ciet. Licet animus ab re abhorreat et a similibus colligendis abalienetur, tamen ut Lector veritatem inspiciat, pauca quaedam ex infinitis, quibus scatet liber iste, ineptiis decerpam. Sane me mei, qui stercora, dum alii e stercoribus flores colligant, quando iam ita necessum est, legam. Et ne longe a principio libri abeamus . . . ex ipsa prima operis pagina exordiamur. . . . ʺ

36. Ibid., 27: "Etrusci non Ethrusci: Etruria non Ethruria: Tusci non Thusci."

37. Ibid., 12: "ex diversis antiquorum monumentis et Hebraicis et Hetruscis" (quoting Curzio); 87: "Hetruscis hisce schedulis"; see chapter VIII. Allacci attacks this orthography, ibid., 27, without explicitly noting its origins with Annius of Viterbo.

38. Ibid., 192: "Pag. 272. *Chartam non possum habere obsidione urgente* . . . in teterrimis enim Scarithis non tantùm chartae scriptae, sed multae etiam non scriptae, quasi scripturarum involucra, quas ipse etiam vidi, in quibus aptissimè scribi poterat, reperiuntur. Quomodo ergo chartam non habuit, qui tam multa chartarum frusta, nullo exarato charactere, sepeliit? Chartam deficere in obsidionibus quis credet?"

39. Ibid., 17: "qui barbarie, ac vulgaribus formis undique scatet."

40. Ibid., 25: "Piget plura persequi in tam humili, et lutulenta scriptione." Allacci's invented word *lutulenta* plays (using the word *lutum*, "mud") on *luculenta*, "brilliant." The alliteration at the beginning of the sentence is, needless to say, deliberate as well.

41. Ibid., 28–29: "In Chrestomathijs hisce colligendis diuturna mora callum iam deduxi stomacho meo, non ita tamen, ut non nauteam bibere malim, quam eas audire, a quibus tamquam ab angue abhorreo. Verborum enim illa portenta, an monstra sunt? Scarith, Caris, Mor, Turg, Asgaria, Vlerda, Dorchethes, Lartes, Saph, Roith, Ochincres, Brocon, Spugi, Barconictus, Ancironae, Schilia, Cronuethia, Schesia, Procravia, Ocalia, Dantelia, Bentia, Porachal, Balth, Rebalth, Rurerebalth, Vosgaria, Onebrae, Enebrae, Inurnes." "Lartes" is the one real Etruscan word.

42. Ibid., 89: "credat qui volet in rebus abditissimis plus vidisse Fesulanum nebulonem, quam alios Prophetas, Deo dilectos, et ad id muneris ab eodem electos, et clariora enunciasse"; 90: "Quod etiam si Fesulanus iste Balaamica asina grandiùs instinctu divino rudeat, haud tamen fieri potest, ut Regem Iudaeorum et nunciatum, et crucifixum proclamet, sed, quemadmodum alii eiusdem haeresis homines, quos sancti Patres pro roborandi fidei veritate adversus gentes afferebant; neque ita sonora, neque ita intelligibili voce enunciarat, sed in re certissima veluti peregrinus balbutiat, et musset."

43. Ibid., 23: "puerilis verborum collocatio in hoc orationis membro"; 25: "Dii te perdant, imperite"; 33: "O Scribam sacri collegii, o plane puerum! O scurram inficetum."

44. Ibid., 202: "Ego Antiquitates hasce, cum charta, pice, aliisque additis longe Thoma recentiores iudico: et utinam non recentissime sint! Equidem praesagit animus, Curtium Inghiramium, qui Fragmenta similia in lucem emisit, hominem catum, prudentem, et verae antiquitatis non subdolae neque efferatae studiosum. . . ."

45. Ibid., a ii v: "Superiori saeculo *Annaeus* quidam *Viterbiensis* inaudita hactenus fraude, ut compararet sibi apud Posteros famam et gloriam ausus est nescio quae ingenii sui lascivientis deliria Berosi antiquioris scriptoris Chaldaei nomine evulgare"; a iii v: "Cum subolfecisset dolum *Leo Allatius* . . . temperare sibi non potuit quin detracta larva et remoto fuco tam insignem fraudem omnibus patefaceret. . . ."

46. The Roman edition was published in 1642 by Vitale Mascardi, together with an additional diatribe against the forger Alfonso Ceccarelli.

Chapter Five

1. Curzio Inghirami, letter to Lucas Holstenius, 23 September 1641, BAV MS Barb. Lat. 6499, 117r:

Molto Illustre Signore e Padrone mio osservantissimo

Hieri nella nostra Accademia de' Sepolti fu trovato nell'urna solita delle compositioni un'involto di lettere stampate sopra il libro del Signor Leone Allacci; procurai d'haverne alcune, delle quali ne mando una à Vostra Signoria per havere con questa occasione di nuovo riverirla. Al Signor Padre [sc. Inghiramo Inghirami], che sperava d'haver havuta sicura promessa di Vostra Signoria della sua venuta qua, gli è stato di disgusto il suo impedimento, et anche di questi Signori che molto la desideravano: et io non nego di non essere tornato di costà suntento [sic; here translated as if the phrase is "su intento"] non havendo potuto ricevere questa gratia; pure spero nella cortesia di Vostra Signoria di essere ancora compiaci-

uto, et in tanto ricordandomele servitore devotissimo con tutto il cuore le bacio le mani, e la riverisco.

> *Volterra li 23 di settembre 1641*
> *Di Vostra Signoria molto Illustre*
> *Devotissimo Servitore Vero*
> *Curtio Inghirami*

2. Lo Spento Accademico Sepolto, *Lettera Sopra il libro intitolato Leonis Allatii Animadversiones in Ethruscarum Antiquitatum Fragmenta* (Florence: Amadore Massi and Lorenzo Landi, 1641).

3. Ibid., 4–5.

4. Melchior Inchofer, S.J., *Bennonis Durkhundurkii Slavi in Spenti Accademici Sepulti Epistolam, Pro Antiquitatibus Etruscis Inghiramis: Adversus Leonis Allatii, contra easdem Animadversiones, Examen* (Cologne: Georg Genselin, 1642; a false imprint, really Lyon). Again, the spelling of "Etruscis" gives away the author's stance as rigorously philological.

5. Ibid., 121–22: "Par in utrisque ignavia, nisi post inventum, maior etiam in Sepulto inventis addere nescienti: qui cum Latine parum, aut nihil, Toscane autem, ut sic dicam, perquam corruptissime sapit, qua lingua in posterum usurus sit, dubitari potest. Aut opinor prorsus (cum honore) tacebit, aut si pruriti respondendi provocetur, antique Etrusce, plane ad noemam, Antiquitatum Inghiramiarum, in grandia mysteria, linguam et calamum exacturus est, fabulando propius crede, quam fando: felicem enimvero tum Inghiramium, opportune scilicet nactum socium, qui cum novis hisce nec unquam veris historiis, novum quoque genus fandi commentus sit. Et sane praestabat utrique nec Latine, nec Italice conari, sed prisca Etruscorum lingua, quidquid tandem libuisset fingere, pauciores futuri erant in Orbis theatro derisores. Nam de cetero, quod saltem spectat ad Sepultum, tam est Latine, vel Etrusce gnarus, (interfecit Lignyphagus) quam simius ille homo putatus Atheniensis, quem iccirco in mare demersit delphis: exemplo prorsus ad harum Antiquitatum imposturam opportuno, etsi plure in his se offerant simii."

6. Melchior Inchofer and Leone Allacci, *Bennonis Durkhundurkii Slavi in Spenti Accademici Sepulti Epistolam, Pro Antiquitatibus Etruscis Inghiramis: Adversus Leonis Allatii, contra easdem Animadversiones, Examen*, BAV, MS Barb. Lat 3061, cc. 33r–55r.

7. The following material is drawn from Thomas Cerbu, "Melchior Inchofer, 'Un homme fin & rusé,'" in *Largo Campo di Filosofare, Eurosymposium Galileo 2001*, ed. José Mongesinos and Carlos Solís (Las Palms de Gran Canaria: Fundación Canaria Orotava de la Ciencia, 2001), 587–611.

8. Ibid., 591, citing a letter of Leone Allacci to Fabio Chigi, BAV, MS Chigi A.III.59, 177r.

9. See ibid., 592–98.

10. Eric Cochrane's description of Inchofer is devastating: "at least some authorities have suspected none other than Galileo's groveling persecutor of 1633, Melchior Inchofer, as the author of a vitriolic satire [the *Monarchy of the Solipsists*, published in 1645] on the society that soon began to circulate in manuscript." *Florence in the Forgotten Centuries*, 212. Redondi, *Galileo Eretico*, 319–21, puts a different slant on Inchofer's participation at Galileo's hearing of 1633.

11. Fabio Chigi's letter from Cologne is preserved in BAV, MS Barb. Lat. 3051, 511r.

12. Rowland, "Etruscan Inscriptions."

13. See Cerbu, "Melchior Inchofer," 599–600, esp. 599n35.

14. See ibid., 599–600.

15. Thomas Cerbu, private communication.

Chapter Six

1. Giovanni Girolamo Carli, letter from Gubbio to unknown addressee, 24 January 1772, Siena, Biblioteca Comunale degli Intronati, MS C.VII.12.

2. Don Secondo Lancellotti, *Farfalloni de gl'Antichi Historici, Notati dall'Abate Don Secondo Lancellotti da Perugia, Accademico Insensato, Affidato, et Humorista Autore del Hoggidí* (Venice: Giacomo Sarzina, 1636).

3. Inghirami, *Discorso*, 1: "Avendo disposto la provvidenza di quel Dio, che il tutto sà, e che tutto può, che in questi tempi delle viscere de' monti venissero alla luce l'Antichità della Toscana, per le quali risorgono dalla Tomba dell'oblio molte notizie degli Antichi secoli, parevami di commettere troppo grave errore a non pubblicarle per mezzo delle Stampe. Ho soddisfatto perciò alla curiosità del Mondo, et al debito mio: ma non prima vedute per la novità del caso hanno cagionato gravi dissensioni fra Litterati."

4. Thomas Dempster, letter to Cardinal Maffeo Barberini, 1 February 1619, BAV, MS Barb. Lat. 2177, c. 5r: "iam Hetruria censoribus, ut audio, placuit, proxima aestate edenda."

5. Ibid., 11r: "Creditoribus bibliothecam meam reliqui, Hetruriam Regalem pro ducentis summis tradere coactus, assignatis pro carcere aedibus propriis, in itinere sicariis Anglis tentatus, gladio me vendicavi."

6. Thomas Dempster, *De Etruria Regali* (Florence: typis Regiae Celsitudinis apud Joannem Cajetanum Tartinum, & Sanctem-Franchium, 1723–26); Cochrane, *Florence in the Forgotten Centuries*, 386–88.

7. Thomas Dempster, *De Hetruria Regali*, Holkham Hall, MS 501.

8. The *Giornale de' letterati di Italia* (XL [1740], 407–12) noted these differences in an article of 1740; see Nicola Parise, *Dizionario Biografico degli Italiani*, vol. 15 (Rome: Istituto dell Enciclopedia Italiana, 1972), 145–47, s.v. *Buonarroti, Filippo*. The Buonarroti papers are preserved in the Biblioteca Marucelliana, Florence; the Biblioteca Medicea Laurenziana, Florence; and the library at Holkham Hall.

9. See Mauro Cristofani, *La scoperta degli etruschi: Archeologia e antiquaria nel '700* (Rome: Consiglio nazionale delle ricerche, 1983).

10. Allacci, *Animadversiones*, 198–99: "Nugarum nihilominus Auctorem nonnulli suspicati sunt Gulielmum Postellum, virum, et linguarum Orientalium peritum, et [e]ffingendas [historias] aptissimum Id tamen sine lege asseritur. Ille quippe ingenti confidentia ad quidlibet audendum projectus, quid in simili agere facturus fuisset, Libris, quos de regione Etruriae edidit; demonstravit, in

quibus licet et pseudo-Berosus, et fabellas Annianas defendit, mul-taque alia praeter communem opinionem ex penu linguarum Ori-entalium insinuet, numquam tamen eò temeritatis devenit, ut adeo impudenter, effrictaque fronte mendacia tam insulsa sparg-eret, sed prudentia, quam à bonorum Auctorum lectione hauserat, cohibuissetque temeritatem, ac licentia omnia blaterandi com-pressisset; cuius male sanas sententias, atuqe inania effata, si ad Fesulanicas confers, Deus bone! Quam cate illa erunt, quam pru-dentia, quam verecunda! Illa hominem scripsisse diceres, haec elinguam bellvam, insulsumque. Sane Postellus homo nequis-simus, ob malas etiam artes, et violatae fidei reus, ut fertur, se-cundo iudicio condemnatus, relagatusque non ita insaniit, neque unquam adeo stupidè, infortunateque in historia delirasset, sed res a semet prudentiori consilio, et verosimiliori apparatu confictas, licet ineptissimus, et meliorum disciplinarum subsidio, et quod magis est, eloquentia destitutus, ordinasset aptius, disposuisset comptius, et elegantius expressisset."

11. See chapter VII.

12. Giorgio Vasari, *Life of Michelangelo Buonarroti*, 1568, from Giorgio Vasari, *Le Vite de' più eccellenti pittori, scultori ed architettori*, ed. Rosanna Bettarini (Florence: Studio per Edizioni Scelte, 1986), vol. V, 14–15:

> *[Michelagnolo] volentieri se ne tornò a Fiorenza, e ... si messe a fare un Cupido che dormiva, quanto il naturale, e finito, per mezzo di Baldassari del Milanese fu mostro a Pierfrancesco [de' Medici] per cosa bella, che giu-dicatolo il medesimo, gli disse "Se tu lo mettessi sotto terra sono certo che passerebbe per antico; mandandolo a Roma acconcio in maniera che paressi vecchio, e' ne caveresti molto più che a venderlo qui." Dicesi che Michelagnolo l'acconciò in maniera che pareva antico: né è da merav-igliarsene, che'l Milanese lo portassi a Roma e lo sotterrassi in una sua vigna, e poi lo vendessi per antico al Cardinal San Giorgio ducati dugento. Altri dicono che gliene vendé un che faceva per il Milanese, che scrisse a Pierfrancesco che facessi dare a Michelagnolo scudi trenta, dicendo che più del Cupido non aveva avuti, ingannando il Cardinale, Pierfrancesco e*

Michelagnolo; ma inteso poi, da chi aveva visto, che 'l putto era fatto a Fiorenza, tenne modi che seppe il vero per un suo mandato, e fece sì l'agente del Milanese gl'ebbe a rimettere, e riebbe el Cupido ... questa cosa non passò senza biasimo del cardinale San Giorgio, il quale non conoscendo la virtù dell'opera , che consiste nella perfezzione, che tanto suon buone le moderne quanto le antiche, purché sieno eccellenti, essendo più vanità quella di coloro che van più dietro al nome che a' fatti: ché di questa sorte di uomini se n'è trovato d'ogni tempo, che fanno piu conto del parere che dell'essere.

[(Michelangelo) gladly returned to Florence ... and set to work making a sleeping Cupid, life-size, and when he had finished it, he showed it through Baldassare Milanese to Pierfrancesco de' Medici as a fine thing, and when the latter had appraised it, he said, "If you put it underground I'm sure it could pass for an ancient work if you sent it to Rome and treated it to look old, and you'd get much more for it than if you sold it here." It is said that Michelangelo treated it to look old, and there is no reason to be surprised that Milanese took it to Rome and buried it in one of his gardens, and then sold it as an antique to the Cardinal San Giorgio for two hundred ducats. Others say that one of Milanese's agents sold it, and wrote to Pierfrancesco that he should give Michelangelo thirty scudi, saying that this was as much as he could get for the Cupid, thus fooling the Cardinal, Pierfrancesco, and Michelangelo; but when it was known later from people who had seen it before that the Cupid had been made in Florence, (Michelangelo) learned the truth, and made Milanese's agent pay him back, and he got back the Cupid ... these things did not happen without discrediting the Cardinal San Giorgio, who did not know the quality of the work, which consists in its degree of perfection, so that modern works can be as good as ancient so long as they are well made; it is vanity when people go after a name rather than the facts, but this kind of man has been found in every era: who takes more account of appearance than essence.]

Vasari's biographies are always superb reading, although not always factually true.

13. Inghirami, *Discorso*, 53: "E se Michelangelo Buonarroti potè sotterrare una statua, e farla ritrovare tra poco tempo come anticha per manifestare la sua mirabile arte; con quanta maggior facilità sarebbe al Fedra Bibliotecario del Vaticano...."

14. Ibid., 54: "Quanto al dire, che il Fedra supponesse per emulare Frate Giovann[i] Annio, che inventò Beroso, e gli altri, si risponde, che egli non era persona da attendere a simili emulazioni."

15. Allacci, *Animadversiones*, 200: "cum is vir prudentissimus, et bonis litteris praeditus, . . . nunquam ad nugamenta ista animum induxisset; neque . . . adeo effreni audacia ad decipiendos posteros pervicaciter accessisset."

16. Ibid., 202: "Equidem praesagit animus, Curtium Inghiramum, qui Fragmenta similia in lucem emisit, hominem catum, prudentem, et verae antiquitatis non subdolae neque efferatae studiosum, et in ea comparada fortunatum. . . ."

17. Inghirami, *Discorso*, 56–57: ". . . se bene occorse ritrovarsi il primo Scarith da una mia sorella d'anni 13 e da me, gli altri dopo sono stati cercati, zappati e cavati da gran quantità et diversità di persone come altre volte ho detto, e se non si son trovati dove Prospero fu assediato era tutta ne' lor beni. . . ."

18. Ibid., 15–16: "Ma molto meno può dubitarsi, che le scritture siano supposte modernamente da chi le dà fuora, perchè è cosa evidente, che quando si cominciarono a trovare, v'intervenne da principio la Corte Criminale, poi tutta la Città di Volterra; vi sono intervenuti molti di quasi tutti i luoghi di Toscana, e di tutte le principali Città, e luoghi d'Italia, et in effetto quanti hanno havuto vaghezza a dichiarirsi della verità sono stati presenti al ritrovamento di queste memorie, il quale non è seguita in una sol volta, ma in un processo di tempo a poco a poco, onde da 25 di Novembre 1634 fin'ad hora sempre se ne sono trovate, e sempre attualmente se ne trovano. Oltre ciò si sono propalate col testimonio autentico d'un processo pubblico formato con grandissime diligenze del Signor Tommaso Medici, e del Signor Ottaviano Capponi, a ciò delegati dal Serenissimo Gran Duca di Toscana, alla presenza di quasi infiniti oculati testimoni, e ratificato poi d'ordine della medesima Altezza, che mandò a posta a riconoscere il luogo, e vedere cavare, i Signori Mario Guiducci, e Niccolò Arrigheti[,] Gentiluomini Fiorentini, il qual protesto si conserva nell'Archivio della Città di Volterra."

19. Ibid., 16:

Non si trova in altri il fine della supposizione, et impostura, et assai meno può trovarsi in me, poiché non v'ho altro utile che l'havere speso le centinaia de gli scudi, ne altro onore, che il ritrovare anticaglie sepolte, che nella stessa guisa potevano esser ritrovate da qualsiasi vil zappatore: almeno se havessi durata una fatica tale, haverei voluto con essa obbligare qualche persona. Ma quando si potesse dubitare, che io havessi potuto far questo per qualche interesse, essendo date fuori le scritture ritrovate fin all'Aprile del 1636, s'era conseguito quanto si desiderava, a che fine dunque adesso dopo la publicazione di quelle se n'haverebbero a fare tante altre con altrettante fatiche, e spese? E tanto più vedendo, che esse hanno tante contrarietà, e tante opposizioni?

. . . E finalmente concessa ogn'altra cosa essendo io per nascita, e professione di costumi sinceri, né per interesse o fine alcuno si può credere, né si può dare, che io habbia fatto supposizioni, o altra cosa indegna; e non può credere altrimenti chi misuri gli altri con la sua misura.

20. Benedetto Buonmattei, letter to Curzio Inghirami; see chapter IV, n. 28.

21. Ingrid D. Rowland, "Th' United Sense of the Universe: Athanasius Kircher at Piazza Navona," *Memoirs of the American Academy in Rome* 46 (2001): 153–81.

22. Athanasius Kircher, *Obeliscus Pamphilius* (Rome: Lodovico Grignani, 1650), 153: ". . . interpretationes . . . in Anni Viterbiensis Beroso apocrypho, verum cum illa ab ipso Authore confictae sunt, indignos existimo, quibus adducendis, et tempus et charta teratur. Huius farinae est Volaterrani inventarum literarum Hetruscarum interpretatio, quae non ita pridem prodiit."

Chapter Seven

1. Giacomo Devoto, *Tabulae Iguvinae* (Rome: typis Regiae Officinae Typographiae, 1948); Augusto Ancellotti, *Le tavole di Gubbio e la civiltà degli umbri* (Perugia: Edizione Jana, 1996).

2. Allacci, *Animadversiones*, 86–88: "In Eugubina enim characteres licet inversi, [87] merè Latini sunt, et nostris similes ut plurimum,

ut singula examinanti patere potest. At quas Faesulanus appingit, licet in nonnullis videantur similes, si tamen omnes considerentur, diversitate sunt, et, quod non contemnendae considerationis est, Eugubina, Hebraeorum more, à dextra in sinistram scripta est; Fesulanicae à laeva in dextram, Romanorum, Graecorum more egeruntur. . . . Observa ergo in Hetruscis hisce schedulis à laeva parte versus semper aequales esse, litterasque ex eodem puncto moveri omnes, quod in dextera non sit, evidentissimo argumento scriptionem illam non ab alia, quam à laeva incipere. . . . Et ut Faesulanus finem faciat blaterandi, vide ibi notas carminibus praepositas, aliquando etiam iteratas, quibus loquentium nomina notari necesse est, [88] omnes in parte laeva. . . . O praeposterum Faesulani ingenium! . . . Si certa est apud Eugubinas illius Tabulae antiquitas, habent qui in Faesulano irrideant, et nugacitatem tanti impostoris compescant."

3. Curzio Inghirami, *Annali Toscani ne' quali si contengono l'origine e progressi delle Città d'Italia*, Volterra, Biblioteca Comunale Guarnacci, MS XLVI.4.1, c. 6: "Da tali consideratione indotto inclinai a si nobile studio da che a me pareva faticoso d'attendere ala Giurisprudenza, applicai l'animo con mio gusto supra l'antiche memorie che nela patria si conservano: ma ben presto appresi dal esperienza non esser lieve cosa scriver historia, e maneggiarla con lode, conobbi non esser guernito dalla natura di tali doti che intramettere fra gli historici mi potessero, che non havendo coltivato con l'industria, e colla medesima inclinatione naturale che mi portava ad intraprendere volontariamente il nobilissimo mestiere di esser historia, correvo risico evidente d'esser come arrogante del mondo ripreso. Ben m'avidi non haver studio d'eloquenzia, necessaria di stile, ne haver contratto habito alcuno di prudenza civile, ma che per solo peccato d'ingegno più a cagione di porto che di studio a ricercare l'Archivii di Volterra più e trarne le disendenze delle famiglie nobili che Pazzioni historiche havevo passato se non perdevo il tempo. Per tanto mi ritrassi dall'opera, abbandonai l'impresa in modo che fin le note fatte da me tralasciate in tutto si perdettero."

4. Ibid.: "Quando non dirò la fortuna, ma la bontà e providenza di quel Dio che il tutto regge e suavemente il dispone, si compiacque farmi casualmente ritrovare nella forma e modo altrove da me provato e dimostrato non più i frammenti, ma posso hormai dire l'Antichità Toscane, le quali benchè non ancor finite d'estrarse e cavarsi dale viscere della terra, m'han facto miglior animo: l'oppositioni a quele date havendomi forzato por propria mia riputatione alla risposta, m'hanno dato occasione di veder qualche autore e da quelo apprendere non poche notizie dell'historia Toscana."

5. Ibid., c. 7–8: "A questo aggionta la consideratione che Volterra Città la quale per comun parere de Classici si Antichi come Moderni scrittori ha tenuto il primo luogo fra le principali di Toscana, anzi d'Italia non habbia hauto chi habbia scritto expressamente: o almeno chi habbia publicato la di lei Historia[;] E che la simplicità di qual Padre fra Mario Giovanelli che haveva ardito di dare alla stampa alcune poche cose ... con agiugnervi nel resto più tosto racconti di favole di donnicciole che verità cavata da scritture autentiche degli Archivii dandole nome di *Cronistoria dell'antichità e nobiltà di Volterra*: have più tosto detratto a me e derogato alla dignità di quella secondo facevano chiaramente apparire l'autorità de Gran Autori, e l'Originali istrumenti, atti e memorie degli Archivi m'hanno di nuovo invagliato talmente a rintracciarne l'historia che di novo al faticar mi ridussi, pure conoscendo insuperabili le difficoltà proposte et impossibile a me ad arrivare a quei requisiti per ben comporla necessarii si per lo stile si per l'carattere sublime che ricerca l'historia si per gl'ornamenti e ampie descrizioni per i discorsi, cagione dell'ationi che in essa si raccontino, et i fini per i quali le cose che si narrano rivenire sieno fatte, e per le sentenze et insegnamenti che dal Historia doverebbe dare in ogni parte di essa." Curzio refers to Giovanelli, *Cronistoria dell'antichità e nobiltà di Volterra*.

6. E. Solaini, "Il falso 'Estratto del Camerotto di Volterra,'" *Rassegna Volterrana* I (1924): 17–19, and the discussion at the end of this chapter. My thanks to the late Angelo Marrucci for showing me this article and gently breaking the news that Curzio Inghirami

was not only a forger of Etruscan antiquities, but of much else besides.

7. See Batistini, "Il ritratto di Fedra Inghirami," 71–72.

8. Ibid., 71–74, concludes that the version of the portrait in the Isabella Stewart Gardner Museum in Boston was painted on commission for Cavaliere Giulio, perhaps by Baldassare Franceschini.

9. Both Fedra's wandering eye and his cleft chin seem to be dominant traits; they have emerged among recent members of the Inghirami family as well.

10. Borrelli, *Le Saline di Volterra*.

11. Batistini, "Il ritratto di Fedra Inghirami," 71–74.

12. For Annius' searing hatred of Greeks, see E. N. Tigerstedt, "Ioannes Annius and *Graecia mendax*," in *Classical, Medieval, and Renaissance Studies in Honor of Berthold Louis Ullmann*, ed. Charles Henderson (Rome: Edizioni di Storia e Letteratura, 1964), vol. 2, 293–310. For his insanity, see chapter VIII.

13. Donald Taylor, *Thomas Chatterton's Art* (Princeton: Princeton University Press, 1978); Grafton, *Forgers and Critics*, 50–54, 57, 58–59, 68; Louise J. Kaplan, *The Family Romance of the Impostor-Poet Thomas Chatterton* (Berkeley: University of California Press, 1989).

14. See also, *cum grano salis*, Eric Hebborn, *Drawn to Trouble: The Forging of an Artist: An Autobiography* (Edinburgh: Mainstream Press, 1991), bearing in mind that Hebborn was an accomplished liar as well as a forger.

15. Angelo Marrucci, *Vol. III: I personaggi e gli scritti: Dizionario biografico e bibliografico di Volterra*, in *Dizionario di Volterra*, ed. Lelio Lagorio (Pisa:Ospedaletto, 1997), 1056–59, s.v. *Inghirami, Curzio*; Bruno Casini, "I 'Libri d'Oro' della Città di Volterra e San Miniato," *Rassegna Volterrana* LXX (1994): 405, mentions three sons.

16. Borrelli, *Le Saline di Volterra*, 65; Marrucci, *Vol. III: I personaggi e gli scritti*,1056–59, s.v. *Inghirami, Curzio*.

17. Borrelli, *Le Saline di Volterra*, 65; Angelo Marrucci, "I bagni e le moie del volterrano alla metà del XVII secolo," *La comunità di Pomarance* 11 (1997): 28–31, 37–41.

18. Raffaello Maffei, *Trattato delle Moie*, Volterra, Biblioteca Comunale Guarnacci, MS XLVII.3.25. See also Borelli, *Le Saline di Volterra*, 65n85.

19. Marrucci, *Vol. III: I personaggi e gli scritti*, 1056–59.

20. Lodovico Inghirami, "Patriziato e cultura a Volterra in età moderna," *Rassegna Volterrana* 70 (1994): 300.

21. Solaini, "Il falso," 17–19; Inghirami, "Patriziato e cultura," 300.

22. Marie Louise Rodén, *Church Politics in Seventeenth-Century Rome: Cardinal Decio Azzolino, Queen Christina of Sweden and the Squadrone Volante* (Stockholm: Almqvist & Wiksell, 2000).

23. Marrucci, *Vol. III: I personaggi e gli scritti*, 1056–59; Casini, "I 'Libri d'Oro,'" 405; Raffaello Maffei, "Vita del Provveditore Raffaello Maffei," in *Storia Volterrana del Provveditore Raffaello Maffei*, ed. Annibale Cinci (Volterra: Tipografia Sborgi, 1887), LVI: "Dopo la morte di Curzio Inghirami, il Provveditore Maffei sposò Orsola di Ser Claudio Ciupi vedova di lui. Sembra che queste nozze non piacessero ai figli che aveva avuti dall'altra moglie, e forse nemmeno alla famiglia Inghirami. Certamente non fu una bella cosa che sposasse la vedova del suo amico svi[s]cerato. Dalla stessa scritta di nozze si ricava quanto tal matrimonio spiacesse ai Maffei. Tra i patti v'è che Orsola non debba abitare in case Maffei ma in quella dei suoi figli, in compagnia di Francesco Lucci, o Luzi, suo zio materno."

24. Ibid., LVII: "Ma da questo secondo matrimonio di Raffaello non nacquero figli, e non sempre la concordia regnò tra i due coniugi, e le liti e le recriminazione ebbero luogo specialmente a cause di interessi. Anzi alla morte del Provveditore, Orsola, che era di 17 anno più giovane di lui, volò sollecita a terze nozze, e quando morì, i figli la vollero deporre nella tomba gentilizia degli Inghirami."

25. Solaini, "Il falso," 17–19. The original manuscript of the *Estratto del Camerotto* is MS LIII.5.1, Volterra, Biblioteca Comunale Guarnacci.

26. Scipione Ammirato the Younger (really named Cristoforo Bianco), ed., *Istorie fiorentine di Scipione Ammirato con l'aggiunte di Scipione Ammirato il Giovane* (Florence: Amadore Massi, 1647).

27. Maffei's *Storia Volterrana* was published only in 1887: Cinci, ed., *Storia Volterrana del Provveditore Raffaello Maffei.*

Chapter Eight

1. "Lectori. Hoc est exemplar Curtio Inghiramio conscriptum, quo si ab his mendaciis abstinuisset suosque litterarios labores in veri investigatione impendisset, tantam sibi gloriam conciliasset, quantum ex his commentariis sui nominis iacturam fecit." Curzio Inghirami, *Ethruscarum Antiquitatum Fragmenta*, Volterra, Biblioteca Comunale Guarnacci, MS LII.6.1

2. Inghirami, *Annali Toscani*, c. 6: "Da tali consideratione indotto inclinai a si nobile studio da che a me pareva faticoso al faticar mi ridussi, pure conoscendo insuperabili le difficoltà proposte et impossibile a me ad arrivare a quei requisiti per ben comporla necessarii si per lo stile si per l'carrattere sublime che ricerca l'historia si per gl'ornamenti e ampie descrizioni per i discorsi, cagione dell'ationi che in essa si raccontino, et i fini per i quali le cose che si narrano rivenire sieno fatte, e per le sentenze et insegnamenti che dal Historia doverebbe dare in ogni parte di essa."

3. Enrico Fiumi, *Volterra etrusca e romana* (Pisa: Pacini Editore, 1976), 11–24; Stefan Steingräber, *Città e necropoli dell'Etruria; Luoghi segreti e itinerari affascinanti alla riscoperta di un'antica civiltà italica* (Rome: Newton Compton Editori, 1983), 94–104; Sybille Haynes, *Etruscan Civilization: A Cultural History* (Los Angeles: The J. Paul Getty Museum, 2000), 363–74; Camporeale, *Gli Etruschi*, 370–78.

4. William V. Harris, *Rome in Etruria and Umbria* (Oxford: Oxford University Press, 1971), 4–28, "The Historiography of Etruria." For Porsenna at Chiusi and Montepulciano, see Rowland, "*L'Historia Porsennae*"; Rowland, "Il mito di Porsenna: Leggenda e realtà," in *Il Mito nel Rinascimento*, ed. Luisa Rotondi Secchi Tarugi (Milan: Franco Cesati Editore, 1993), 391–407; and Rowland, "Due 'traduzioni' rinascimentali dell'*Historia Porsennae*," in *Protrepticon: Studi in memoria di Giovannangiola Secchi Tarugi*, ed. Sesto Prete (Milan: Istituto di Studi Umanistici Francesco Petrarca, 1989), 125–33. For the brothers Vibenna and Macstrna in Rome, see Jocelyn Penny Small, *Cacus and Marsyas in Etrusco-Roman Legend* (Princeton: Princeton University Press, 1982). For Maecenas, see

Francesco Dini, *Dell'origine, famiglia, patron, e azzioni di Caio Mece-nate gran capitano* (Venice: a spese di D. Lovisa, 1704); Paola Za-marchi Grassi and Dario Bartoli, *Museo archeologico nazionale G. Cilnio Mecenate, Arezzo* (Rome: Istituto poligrafico e Zecca dello Stato, Libreria dello Stato, 1993); Riccardo Avallone, *Mecenate* (Naples: Libreria scientifica editrice, 1963); and Franco Paturzo, *Mecenate, il ministro di Augusto : Politica, filosofia, letteratura nel peri-odo augusteo* (Cortona: Calosci, 1999).

5. Propertius IV.10, 27–30:
   heu Vei veteres! Et vos tum regna fuistis,
   et vestro posita est aurea sella foro:
   nunc intra muros pastoris bucina lenti
   cantat, et in vestris ossibus arva metunt.

6. See chapter II.

7. The copy of the 1512 edition of Annius of Viterbo's *Antiquitates* in the Biblioteca Hertziana in Rome has a series of marginalia in a contemporary hand that declares, inter alia (11v): "This man went insane twice and because of his insanity died in chains; rightly, therefore, in his turn he teaches the art of going insane, and has al-ready attracted not a few to him, so that there is no lack of wit-nesses to his ignorance"; and further laments about Annius' public (168v): "Alas, you true unfortunates! O crazy delirious man—it's no wonder you died in chains!" ["Hic vir bis insanivit et propter in-saniam mortuus est vinctus; recte igitur et suo vice artem docet in-saniendi: et iam nonnullos traxit ad se ut ignorantiae suae non desunt testes." "Heu infelices certe. O hominem amentem delirum nimirum si mortuus es in vinculis."] The story that he was poi-soned by Cesare Borgia is refuted by Tigerstedt, "Ioannes Annius," vol. 2, 299; and Stephens, "Berosus Chaldaeus," 208n74.

8. The "Egyptian stele" is still on view in the Museo Civico of Viterbo; see Adriana Emiliozzi, *Il Museo Civico di Viterbo: Storia delle raccolte archeologiche* (Rome: Consiglio Nazionale delle Ricerche, 1986), 19–35; Stephens, "Berosus Chaldaeus," 166–76; and Roberto Weiss, "An Unknown Epigraphic Tract by Annius of Viterbo," in *Italian Studies Presented to E. R. Vincent*, ed. E. R. Vin-

cent, K. Forster, and U. Limentani (Cambridge: Heffer, 1962), 119.

9. Rowland, "*L'Historia Porsennae*," passim.

10. On Annius and his method, see Stephens, "Berosus Chaldaeus"; Stephens, "The Etruscans and the Ancient Theology in Annius of Viterbo," in *Umanesimo a Roma nel Quattrocento*, ed. Paolo Brezzi and Maristella Panizza Lorch (New York: Barnard College, 1984), 309–22; Christopher Ligota, "Annius of Viterbo and Historical Method," *Journal of the Warburg and Courtauld Institutes* 50 (1987): 44–56; Danielsson, "Annius von Viterbo"; Fumagalli, "Un falso tardo-quattrocentesco"; Grafton, *Forgers and Critics*, 54–68, 104–23; Gigliola Bonucci Caporali, ed., *Annio da Viterbo, Documenti e ricerche* (Rome: Consiglio Nazionale delle Ricerche, 1981), esp. Giovanni Baffioni, "Viterbiae Historiae Epitoma: Opera inedita di Giovanni Nanni da Viterbo," ibid.; and Roberto Weiss, "Traccia per una biografia di Annio da Viterbo," *Italia medioevale e umanistica* 5 (1962): 425–41.

11. Deitz, "Die Scarith von Scornello."

12. Rowland, "*L'Historia Porsennae*."

13. Eric Cochrane gives a synthetic picture of chroniclers and their art in *Historians and Historiography in the Italian Renaissance* (Chicago: University of Chicago Press, 1981), 9–15. His remarks on Curzio Inghirami are curt and hilarious, 443, 591.

14. The *Hypnerotomachia Poliphili* has been edited in Italian by Giovanni Pozzi and Lucia Ciapponi, *Hypnerotomachia Poliphili* (Padua: Antenore, 1980), and more recently by Marco Ariani and Mino Gabriele, *Hypnerotomachia Poliphili* (Rome: Adelphi, 1998). The English translation on CD by Ian White (Octavo: in press) is much more accurate than that in print by Joscelyn Godwin (Thames and Hudson, 2000).

15. Vincenzo Renieri's *Monopanthon, De Ethruscarum Antiquitatum fragmentis Scornelli prope Vulterram repertis Disquisitio Astronomica* was published in 1638 by Amadore Massi in Florence. The text is also published entire in Lisci, *Documenti raccolti*, 105–8.

16. The only compendium of Etruscan texts from manuscripts to date is O. A. Danielsson, *Etruskische Inschriften in Handschriftlicher Über-*

lieferung, Skrifter utgivna av Kungliga Humanistiska Vetenskaps-Samfundet i Uppsala, 25:3 (Uppsala: Almqvist and Wiksell, 1928), to which can now be added Mauro Cristofani, "Le iscrizioni etrusche," in *Siena: Le Origini: Testimonianze e miti archeologici*, ed. Mauro Cristofani (Florence: Leo S. Olschki, 1979), 119–25; Marina Martelli Cristofani, "MS Sloane 3524," in Cristofani, *Siena: Le Origini*, 136–43; Rowland, "Etruscan Inscriptions," where, like Chigi, I fail to note the Clusine "H" in the third name of Larthi Seianti Hanunia Crapilusa. In a forthcoming *History of Etruscan Studies, 1450–1750*, I will provide a comprehensive analytical list with correct readings.

17. Massa-Pairault, "La stele di 'Avile Tite.'"

18. Raffele Maffei, *Commentaria Urbana* (Rome: Eucharius Silber, 1506), CCCCLXIIr.

19. Olaus Magnus, *Historia de gentibus septentrionalibus, earumque diversis statibus, conditionibus, moribus, ritibus, superstitionibus, disciplinis . . . Authore Olao Magno Gotho* (Rome: apud Ioannem Mariam de Viottis parmensem, 1555), was already translated into Tuscan vernacular by 1565 (*Historia delle genti et della natura delle cose settentrionali da Olao Magno . . . descritta in xii libri. Nuovamente tradotta in lingua toscana* [Venice: Appresso i Giunti, 1565]). Johannesson, *Renaissance of the Goths*.

20. See especially Vulcanius, *De literis et lingua Getarum*.

21. Curzio Inghirami, *L'Amico Infido*, Volterra, Biblioteca Comunale Guarnacci, MS LIII.4.14, c. 26: "biduo hic manere decrevi per vedere le pulcherrime scritture, che audivi nuper esser ritrovate mirabile modo in quadam Villa qui propinqua. Ho inteso che in his continentur omnia antiquissima memoranda di tutta l' Europa, le quali antehac omnibus fuere ignota." See also ibid., 26: "Pedante: questa antiquissima et vetustissima Città, quae olim omnis Hetruriae erat caput."

22. Inghirami, *Ethruscarum Antiquitatum Fragmenta*, †2r.

23. Ibid., 272: Scarith 87: "Levavit vocem vultur à facie Locustae. Locusta Leones vorabit. Lapides horrore sudabunt."

24. André Rochon, *Formes et significations de la "Beffa" dans la littérature*

italienne de la Renaissance: Études réunies (Paris: Université de la Sorbonne Nouvelle, 1972–75).

25. Jane Whitehead suggested a connection between the two in "Laughing Out Loud: A Study in Etruscan and Renaissance Humor," paper presented at the annual meeting of the Renaissance Society of America, Florence, Italy, March 23, 2000.

26. Inghirami, *Ethruscarum Antiqtuitatum Fragmenta*, 7. Ernst and Gaudenzio already noted the parallels, *Ad antiquitates etruscas*, 61; Deitz, "Die Scarith von Scornello," also provides some telling parallels.

27. Annius of Viterbo, *Commentaria Fratris Joannis Annii Viterbiensis super Opera Diversorum Auctorum de Antiquitatibus Loquentium* (Rome: Eucharius Silber, 1498); Stephens, "Berosus Chaldaeus," provides the most detailed account of the complicated way in which Annius constructs this argument.

28. Inghirami, *Ethruscarum Antiqtuitatum Fragmenta*, 9, 12.

29. Rowland, "L'Historia Porsennae," 188, 192.

30. Pallottino, *Testimonia Linguae Etruscae*, 843; cf. Pallottino, *Thesaurus Linguae Etruscae*, 416, citing Servius, *Commentary on Vergil's Aeneid*, II.278; VIII.475. See also Mauro Cristofani, *Dizionario della civiltà etrusca* (Florence: Giunti Martello, 1985), 160, s.v. *lucumone*.

31. For the demotion of Lars Porsenna (here called Laertes), see Inghirami, *Ethruscarum Antiquitatum Fragmenta*, 273.

32. See chapter V.

33. See note 10 *supra*.

34. Allacci, *Animadversiones*, 20: "[Scriptores Etruscarum antiquitatum] semper enerves, languidi, concisi, incompti, balbi, nisi ubi garrulitas locum habet, barbari: semper timidi ne errent, nullo antiquitatis, nullo vetustatis succo pingues, aut conspicui . . ."; 25: "Dii [*sic*] te pereant, imperite. Plebeia locutio, et e faece vulgi."

35. Inghirami, *Discorso*, 43: "È opposto primieramente a' libri de gli autori supposti da Annio, che essi siano scritti col medesmo stile de' Commentari del medesimo Annio; il che non si può dire dell'Antichità Toscane ritrovate, e date in luce senza commento, o aggiunto d'alcuno."

36. At least one of the authors cited by Annius of Viterbo, Metas-
thenes the Persian, is entirely fictitious; see Stephens, "Berosus
Chaldaeus," 43–44. Inghirami, *Discorso*, 54: "Annio suppose Autori,
che sapevasi avere scritto: onde questi non doveva inventare Autori non più uditi, che si poteva imaginare, che non sarebbero stati
creduti, ne stimati, non mancandogli scrittori antichi da potergli
attribuire la supposizione i quali più verisimilmente potevano
esser creduti veri"; 55: "Annio tradusse, e comentò quello che suppose, e forse furno principali cagione della supposizione non tanto
l'invenzioni, e fallacie scritte ne suoi Autori quanto per poterli co-
mentare a suo modo, et anche mostrare la sua erudizione, ma
questi ne pur vi mette una sillaba. Annio publicò, e dette in luce
quello che suppose; questi averebbe risposto la supposizione dove
non si potesse per ragione umana mai più ritrovare."

37. Inghirami, *Discorso*, 55: "Il fine d'Annio fu di ridurre a Viterbo tutta
quella grandezza, et antichità, che all'altre Città si perveniva ma
il fine di chi havesse supposte l'Antichità Toscane sarebbe stato
in tutto a quello d'Annio dissimile, dicendosi in esse di
ciascheduna Città indifferentemente le lodi convenienti non dis-
cordando da quello che ne dichino gli Autori Classici . . . qual-
sivoglia concluderà, che queste scritture non posso essere state in-
ventate dal Fedra, né da altri per emulare le finzioni d'Annio."

38. Ibid., 18: "E gli originali si son mostrati, e si mostrano continua-
mente a chi li vuol vedere."

39. Giovanni Girolamo Carli in Gubbio, letter to an unknown addressee, 24 January 1772, Siena, Biblioteca Comunale degli Intronati, MS C.VII.12, c. 44r–v:

*Il fu Ippolito Cigna, buon Pittore, e Poeta Volterrano, ma oriundo di Colle
di Val d'Elsa. . . . Mi disse adunque . . . che egli avendo avuto sotto gli
occhi uno degli* Scariti *originali, ci era chiarito, che i caratteri minuscoli
del medesimo sono di quella precisa forma particolare, che era in uso verso
il fine del sec. XV, vale a dire al tempo di Fedra, e la Carta è absolutamente
della fabbrica delle Cartiere di Colle, scorgendosi perfino il marco di una
di quelle Cartiere simile ad altro da lui osservato in un de' Libri delle* Pro-

visioni *del Pubblico di Colle dello stesso sec. XV esistenti nell'Archivio di essa Città.*

[The late Ippolito Cigno, a good Painter and Poet from Volterra, but born in Colle di Val d'Elsa . . . told me . . . that once he had one of the original Scarith before his eyes, and he decided that the lowercase letters of the same were of the same particular form that was in use toward the end of the fifteenth century, that is, at the time of Fedra, and the Paper is absolutely of the same composition as that from the Factories at Colle; he even made out the mark of one of those Factories, similar to one he observed in one of the Registers of the City of Colle, preserved in the Archives of that City, and from the fifteenth century.]

Had Ippolito Cigna also had a sample of Curzio Inghirami's handwriting, he would have reached another conclusion about the handwriting of the scarith.

Chapter Nine

1. Ranuccio Bianchi Bandinelli, "Murlo (Siena) — Monumenti archeologici nel territorio," *Notizie degli Scavi* 5 (1936): 165–70.

2. Kyle M. Phillips Jr., *In the Hills of Tuscany: Recent Excavations at the Site of Poggio Civitate* (Philadelphia: University Museum, University of Pennsylvania, 1993); Kyle M. Phillips Jr. and Erik O. Nielsen, "Poggio Civitate," in *Case e Palazzi d'Etruria*, ed. Simonetta Stopponi (Milan: Electa, 1985), 64–69.

3. Ingrid E. M. Edlund-Berry, *The Seated and Standing Statue Akroteria from Poggio Civitate (Murlo)* (Rome: Giorgio Bretschneider, 1992).

4. Ingrid D. Rowland, "Early Attestations of the Name 'Poggio Civitate,'" in *Murlo and the Etruscans: Art and Society in Ancient Etruria*, ed. Richard D. De Puma and Jocelyn Penny Small (Madison: University of Wisconsin Press, 1994), 3–5.

5. Nancy Thomson de Grummond, "Rediscovery," in *Etruscan Life and Afterlife*, ed. Larissa Bonfante (Detroit: Wayne State University Press, 1986), 18–46; Cipriani, *Il mito etrusco*; Rowland, "*L'Historia Porsennae.*"

6. Ranuccio Bianchi Bandinelli, *Dal diario di un borghese: Con i diari inediti (1961–1974)* (Rome: Editori riuniti, 1996); obituaries of Kyle M. Phillips Jr. by Richard De Puma and Jocelyn Penny Small in their *Murlo and the Etruscans*, xxvi–xxviii; by Richard De Puma, Ingrid E. M. Edlund-Berry, and Lucy Shoe Meritt in *American Journal of Archaeology* 93 (1989): 239–40; by Erik O. Nielsen in *Studi Etruschi* 57 (1991): 485–90.

7. Interestingly, Enrico Fiumi proposed looking more closely at Poggio Civitate after studying the results of Kyle Phillips's first excavation in nearby Spannocchia, where a series of Etruscan tombs showed strong Volterran connections; as Fiumi wrote, Phillips was probably just breaking ground at Poggio Civitate, Enrico Fiumi, *I confini della Diocesi Ecclesiastica del Municipio Romano e dello Stato Etrusco di Volterra* (Florence: Leo S. Olschki, 1968), 58–59. Kyle Phillips told me the story about "digging in the bushes" on several occasions in the late 1970s and early '80s.

8. Phillips and Nielsen, "Poggio Civitate," 65; see also Giovanni Colonna, "Ricerche sull'Etruria interna volsiniese," *Studi Etruschi* 41 (1973): 45–72, esp. 72.

9. Ingrid E. M. Edlund-Berry, *The Gods and the Place: Location and Function of Sanctuaries in the Countryside of Etruria and Magna Graecia (700–400 B.C.)* (Stockholm: Svenska institutet i Rom, 1987).

10. Dionysius of Halicarnassus, *History*, 3.51, a league against Rome formed by Arezzo, Chiusi, Roselle, and Vetulonia.

11. Phillips and Nielsen, "Poggio Civitate," 65; Phillips, *In the Hills of Tuscany*; Ingrid E. M. Edlund-Berry, "Ritual Destruction of Cities and Sanctuaries: The 'Un-founding' of the Archaic Monumental Building at Poggio Civitate (Murlo)," in De Puma and Small, *Murlo and the Etruscans*, 16–28.

12. This viewpoint is put most forcefully in Comune di Bologna, Museo Civico Archaeologico, *Principi Etruschi tra Mediterraneo ed Europa* (Venice: Marsilio, 2000), which devotes a whole section to "il palazzo del principe." The words *palazzo* and *principe* grow out of imperial Rome, from the time of the *princeps* Augustus and his residence on the Palatine Hill (the first *palatium*); the Holy

Roman Emperor, the pope, Renaissance warlords like Sigisi-
mondo Malatesta or Federico da Montefeltro, and tyrants like
Lorenzo de' Medici adopted this vocabulary as a means of self-
aggrandizement in the classical style. The connection of this set of
concepts with the Etruscans, therefore, is a tortuous rather than
a natural one, the equivalent of Leonardo Dati's chivalric *Gesta
Porsemnae Regis*.

13. Phillips and Nielsen, "Poggio Civitate," 68; Ingrid D. Rowland,
    "Etruscan Secrets," *New York Review of Books*, July 5, 2001.

14. Inghirami, *Ethruscarum Antiquitatum Fragmenta*, ††ii 2v.

15. Enrico Fiumi, the great Volterran archaeologist, called fibulae of
    the kind that Curzio discovered "a navicella," "like a little ship."
    Volterra Etrusca e Romana, 33–36, which includes a catalog of the ob-
    jects on display in the Museo Guarnacci, Volterra; the series of
    bronze fibulae he describes in Case VI includes several incised ex-
    amples that provide extremely close parallels to Curzio's find.
    Ibid., 36. The more common term now for this type of fibula is the
    less glamorous "a sanguisuga" "like a bloodsucker."

16. Ibid.; Stefan Steingräber and Horst Blanck, eds., *Volterra: Etrus-
    kische un mittelalterliches Juwel im Herzen der Toskana* (Mainz: Philipp
    von Zabern, 2003); Steingräber, *Città e necropoli dell'Etruria*,94–116,
    "Volterra e l'Etruria nord-occidentale"; Gabriele Cateni, *L'acropoli
    di Volterra: Nascita e sviluppo di una città* (Pisa: Pacini, 1981); Ernesto
    Galluccio, *Volterra etrusca alla luce delle nuove scoperte. Opuscula Ro-
    mana: Annual of the Swedish Institute in Rome*, 2000.

17. Borrelli, *Le Saline di Volterra*; Marrucci, "Nützliche Metalle."

18. For the siege of Volterra, see Fiumi, *L'impresa di Lorenzo de' Medici*;
    its implications for Le Moie are discussed by Borrelli, *Le Saline di
    Volterra*, 14. For the office of Provveditore del Sale, see Borrelli,
    Le Saline di Volterra, 58.

19. See Andrea Augenti and Monica Baldassarri, "Das Kastell Monte
    Voltraio: Geschichte und Archäologie," in *Otto der Große und Eu-
    ropa: Volterra von Otto I bis zur Stadtrepublik*, ed. Andrea Augenti
    (Siena: Nuova Immagine, 2001), 46–50; and Andrea Augenti, "Die
    Siedlungsstrukturen des Territoriums," in ibid., 39–45.

20. See note 32.

21. Caciagli, *La casa colonica.*

22. *Processo del Ritrovamento delle Scritture Antiche di Scornello*, Volterra, Biblioteca Comunale Guarnacci, LII.6.5, c. 4: "di presente in detto luogo non è altro che una Casa per servitio del Padrone, et altra per servitio de contadini et Mezzaioli . . . da cento anni sono nello stato che è di presente anzi più macchioso et salvatico."

23. Ibid., c. 5v: "Come il di 13 Dicembre fu cercato con maggiore diligenza et ritrovato sotto un muro molto forte che andava di Mezzo giorno a levante circa due braccia sotto terra, sotto una gran Pietra che pareva Calcestruzzo fra due pietre Bianche in forma di Urna, una Mestura più grande della prima, et nella quale erano alcune lettere Etrusche che per l'antichità non si poterono conoscere. . . ."

24. Ibid.: "Come detto luogo davanti si cominciassi a cercare non mostrava se non macie di sassi, macchie et sterpi non conoscendosi muri sopra terra se non in due o tre parti."

25. Ibid.: "Come il di 29 del detto Mese si cercò con più diligenza cominicandosi da Mezzogiorno verso dove s'era trovata la prima scrittura e cercando in detto luogo si trovò muraglie grosse e forti fatti a calcina, volte franate, et altre rovine."

26. Ibid., c. 5v–6r: "Come fra le altre rovine et Muri fondamentali rimasti [6r] in piedi erano macchie, cioè Quercie, lentischi, e sondri con barbe grosse che mostrano molta antichità e anco vi si sono trovati Ceppi grossi fradici per esser stati tagliati più volte et non rimessi."

27. Ibid., c. 6r: "Come le rovine di Tevoli rotti, Mattoni, et sassi et muraglie anco data la volta erano nel modo che li ha portato il caso, onde alla prima rovina in quà non par possibile che vi sia stati cercato, cavato, e messo."

28. Ibid.: "Come in più luoghi si è trovato Ceneri, et carboni, benche fradici, et ossa humane messe bruciate ogni cosa confusa con le Rovine. . . ."

29. Ibid., c. 6r–v: "Come il 30 Dicembre da mattina nel ultimo fondamento d'un Muro intorno a due bracchia e mezzo sotto terra si trovò una Mestura simile la scrittura enunciata sotto No. 3."

30. Ibid., cc. 6v–7r: "Come il di 5 febbraio lontano al luogo detto di sopra passi sette antichi Romani alla fine del detto Muro anda[n]te verso tramontana [7r] fu trovato nella terra una statuetta di stagno con lettere Etrusche nel'orlo della veste, et uno stromento di Bronzo, et sopra nel muro murato una stagnata nella quale fu Mestura di pece et rinvolti di care vi era le scritture enunciate sotto numero 8."

31. Ibid., c. 7r: "Come nel cercar detto luogo si è avvertito e visto s si vede come gia vi era una Rocca di forma rotonda di braccia 200 che anche verso levante haveva congionto altro edifitio."

32. Ibid., c. 10r: "fu visto essere un monte alle radici del quale verso levante è la Zambra fiumicello che entra in Cecina, la quale scorre da mezzo girono à Ponente, et entra in Mare. Da Ponente verso tramontana sono più edifizii delle Moie e Saline, e da questa parte la Salita al Monte è assai facile, ma dalla parte della Zambra è molto scosceso e difficile. Da settentrione circa tre miglia e mezzo lontana è la Città di Volterra, e da quattrocento braccia discosto pure dall'istessa parte e la casa della Villa delo Signore Inghiramo Inghirami cioè l'habitatione per il padrone, e per i contadini. Vicino a detto luogo è la strada che conduce al fiume Cecina, e alle Pomarance. Fu dunque considerato molto bene il paese da sopradetti Signori Deputati e fu visto esser luogo di pastura, e selvaggio, e pieno di macchie e di sterpi, e d'alberi parte de quali mediante la deradicazione fatta di loro per cercare denotano per le loro barbe molto grosse, e loro antichi ceppi essere di tempo antichissimo nati in detto luogo. Si veddero inoltre molte vestigie di muraglie antiche, delle quali solamente sopreterra n'apparivano alcuni vestigii una dalla parte di Mezzo giorno vero il fiume Cecina, e l'altra da parte del Ponente furono le suddette muraglie fatte riconoscere dal Messer Francesco Fantoni uno degli' Ingegneri del Magistrato de' Signori Capitani di Parte di Firenze, e da Mastro Giovanni Maria Sanfinochii Volterrano s si venne in cognition essere dentro ad un recinto di mural tondo di circuito braccia 200 fiorentine, al quale recinto denota essere stata una Rocca. Si fecero parimente ricercare alcuni vestigie di muraglie contigue alla Rocca

onde si conobbe essersi stati gia altri edifizii congiunti e si considerorno anco altre vestigie in diversi luoghi vicini con parte di grosse mura, che si vedde essere le mura Castellane, e si venne in chiaro esservi stato un Castello di braccia 3207 fiorentine di circuito. Dentro il recinto delle cui mura si scorgono molte rovine, e muraglie di antichissimo tempo, onde del tutto fu ordinato al medesimo Fantoni che con ogni diligenza ne levi la pianta."

33. See Caciagli, *La casa colonica*, 348.

34. See ibid., 83–84.

35. No explicit site survey seems to have been undertaken in the region of Scornello. The discrediting of the scarith has proven nearly as effective as the ritual destruction of Poggio Civitate. However, there has been a good deal of work in the region of Volterra and the Cecina valley; see Andrea Augenti and Nicola Terrenato, "Le sedi del potere nel territorio di Volterra: Una lunga prospettiva (secoli VII a.C.–XIII d.C.), in *II. Congresso nazionale di archeologia medievale: Musei civici, Chiesa di Santa Giulia, Brescia, 28 settembre–1 ottobre 2000*, ed. Giovanni Pietro Brogiolo (Florence: All'Insegna del Giglio, 2000), 298–303; Nicola Terrenato and Alessandra Saggin, "Ricognizioni archeologiche nel territorio di Volterra," *Archeologia Classica* 46 (1994): 465–82; Paolo Carafa, "Organizzazione territoriale e sfruttamento delle risorse economiche nell'agro Volterrano tra l'orientalizzante e l'età ellenistica," *Studi Etruschi* 59 (1994): 104–21, as well as the earlier work of Fiumi, *I confini della Diocesi*. Guglielmo Maetzke and Luisa Tamagno Perna, eds., *Aspetti della cultura di Volterra etrusca fra l'età del ferro e l'età ellenistica e contributi alla ricerca antropologica alla conoscenza del popolo etrusco, Atti del Convegno di Studi Etruschi ed Italici, Volterra, 15–19 ottobre 1995* (Florence: Olschki, 1997).

36. See the cadastaral map of 1819 in Caciagli, *La casa colonica*, 348. The castle of Zambra mentioned in Ottonian records certainly brings the Torrente Zambra to mind, and Andrea Augenti locates this citadel near Scornello in the map that accompanies his "Die Siedlungsstrukturen," 44. Nearer Pisa, on the other hand, another

Torrente Zambra passes beneath the medieval Castello di Calci and the seventh-century Pieve of San Jacopo in Zambra, still an active parish; see their website at http://www.parrocchie.it/cascina/zambra/.

37. Rowland, "*L'Historia Porsennae*," 188, 192.

38. See above, note 32.

39. Steingräber, *Città e necropoli dell'Etruria,* 337–38.

Bibliography

Manuscripts

Allacci, Leone. Letters to Fabio Chigi. BAV, MS Chigi A.III.59.

Buonmattei, Benedetto. Collected Papers. Florence, Biblioteca Nazionale Centrale, MS Magliabechi IX.122.

Carli, Giovanni Girolamo. Letter from Gubbio to an unknown addressee, 24 January 1772. Siena, Biblioteca Comunale degli Intronati, MS C.VII.12, 42r–45v.

Chigi, Fabio. Letter from Cologne. BAV, MS Barb. Lat. 3051, 511r.

Dempster, Thomas. *De Hetruria Regali*. Holkham Hall, MS 501.

——. Letters to Cardinal Maffeo Barberini, BAV, MS Barb. Lat. 2177.

Gaudenzio, Paganino. *De Charta: Exercitatio*. BAV, MS Urb. Lat. 1605, 73r–103v.

——. *Epithalamium* for Grand Duke Ferdinando and Vittoria della Rovere. BAV, MS Urb. Lat. 1580, 2r.

——. *Paganini Gaudentii in Antiquitates quasdam editas sub nomine Prosperi Faesulani Animadversio*. BAV, MS Urb. Lat. 1605, 104r–130r.

Inchofer, Melchior, and Leone Allacci. *Bennonis Durkhundurkii Slavi in Spenti Accademici Sepulti Epistolam, Pro Antiquitatibus Etruscis Inghiramis: Adversus Leonis Allatii, contra easdem Animadversiones, Examen.* BAV, MS Barb. Lat 3061, cc. 33r–55r.

Inghirami, Cavaliere Giulio. Letters to Lord Bailiff Andrea Cioli, 29 March 1635. BAV, MS Barb. Lat. 3150, c. 264r–265r.

Inghirami, Curzio. *L'Amico Infido.* Volterra, Biblioteca Comunale Guarnacci, MS LIII.4.14.

———. *Annali Toscani ne' quali si contengono l'origine e progressi delle Città d'Italia.* Volterra, Biblioteca Comunale Guarnacci, MS LIII.5.1.

———. *Estratto delle Scritture, che esistono nell'Archivio ossia Camerotto della Città di Volterra fatto nel 1562.* Volterra, Biblioteca Comunale Guarnacci, MS LIII.5.1.

———. *Ethruscarum Antiquitatum Fragmenta*, autograph. Volterra, Biblioteca Comunale Guarnacci, MS LII.6.1.

———. Letters. Volterra, Archivio Inghirami, filza 59.

———. Letter to Lucas Holstenius, 23 September 1641. BAV MS Barb. Lat. 6499, 117r.

Maffei, Raffaello. *Trattato delle Moie.* Volterra, Biblioteca Comunale Guarnacci, MS XLVII.3.25.

Noghera, Vincenzo. Letters to Cassiano dal Pozzo. Rome, Biblioteca Corsiniana, Archivio dal Pozzo, MS. XII (10).

———. Letters to Lucas Holstenius and Cardinal Francesco Barberini. BAV, MS Barb. Lat. 6472.

———. Letter to Cardinal Giulio Sacchetti, copy for Fabio Chigi. BAV, MS Chigi G.II.65, 231r–237v.

Processo del Ritrovamento delle Scritture Antiche di Scornello. Volterra, Biblioteca Comunale Guarnacci, MS LII.6.5.

Books and Articles

Allacci, Leone. *Apes urbanae, sive de viris illustribus, qui ab anno MDCXXX per totum MDCXXXII Romae adfuerunt ac typis aliquid evulgarunt.* Rome: Ludovico Grignani, 1633.

———. *Leonis Allatii Animadversiones in Antiquitatum etruscarum fragmenta ab Inghiramio edita*. Paris: S. Cramoisy, 1640.

———. *Socratis, Antisthenis, et aliorum socraticorum epistolae. Leo Allatius hactenus non editas primus graece vulgavit; latine vertit; notas adiecit; dialogum, De scriptis Socratis, praefixit*. Paris: S. Cramoisy, 1637.

Ancellotti, Augusto. *Le tavole di Gubbio e la civiltà degli umbri*. Perugia: Edizione Jana, 1996.

Annius of Viterbo, *Commentaria Fratris Joannis Annii Viterbiensis super Opera Diversorum Auctorum de Antiquitatibus Loquentium*. Rome: Eucharius Silber, 1498.

Ariani, Marco, and Mino Gabriele. *Hypnerotomachia Poliphili*. Rome: Adelphi, 1998.

Asor Rosa, Alberto, ed. *Letteratura italiana*. Vol. 2: *Produzione e consumo*. Torino: Einaudi, 1983.

Augenti, Andrea. "Die Siedlungsstrukturen des Territoriums." In *Otto der Große und Europa: Volterra von Otto I bis zur Stadtrepublik*, ed. Andrea Augenti, 39–45. Siena: Nuova Immagine, 2001.

———, ed. *Otto der Große und Europa: Volterra von Otto I bis zur Stadtrepublik*. Siena: Nuova Immagine, 2001.

Augenti, Andrea, and Monica Baldassarri. "Das Kastell Monte Voltraio: Geschichte und Archäologie." In *Otto der Große und Europa: Volterra von Otto I bis zur Stadtrepublik*, ed. Andrea Augenti, 46–50. Siena: Nuova Immagine, 2001.

Augenti, Andrea, and Nicola Terrenato. "Le sedi del potere nel territorio di Volterra: Una lunga prospettiva (secoli VII a.C.–XIII d.C.). In *II. Congresso nazionale di archeologia medievale: Musei civici, Chiesa di Santa Giulia, Brescia, 28 settembre–1 ottobre 2000*, ed. Giovanni Pietro Brogiolo, 298–303. Florence: All'Insegna del Giglio, 2000.

Avallone, Riccardo. *Mecenate*. Naples: Libreria scientifica editrice, 1963.

Batistini, Giovanni. "Il ritratto di Fedra Inghirami, opera di Raffaello: Il vero e il falso." *Rassegna Volterrana* 71–72 (1994–95): 59–75.

Battistini, Mario. *L'ammiraglio Jacopo Inghirami e le imprese dei cavalieri dell'Ordine di S. Stefano contro i Turchi nel 1600.* Volterra: Tip. Confortini, 1912.

Bavoni, Umberto. *La Cattedrale di Santa Maria Assunta e il Museo Diocesano di Arte Sacra di Volterra.* Florence: Edizioni IFI, 1997.

Biagioli, Mario. *Galileo Courtier.* Chicago: University of Chicago Press, 1993.

Bianchi Bandinelli, Ranuccio. *Dal diario di un borghese: Con i diari inediti (1961–1974).* Rome: Editori riuniti, 1996.

———. "Murlo (Siena)—Monumenti archeologici nel territorio." *Notizie degli Scavi* 5 (1936): 165–70.

Blackwell, Richard. *Galileo, Bellarmine, and the Bible.* Notre Dame: University of Notre Dame Press, 1992.

Bologna, Museo Civico Archaeologico. *Principi Etruschi tra Mediterraneo ed Europa.* Venice: Marsilio, 2000.

Bonfante, Larissa, ed. *Etruscan Life and Afterlife.* Detroit: Wayne State University Press, 1986.

Bonucci Caporali, Gigliola, ed. *Annio da Viterbo, Documenti e ricerche.* Rome: Consiglio Nazionale delle Ricerche, 1981.

Borrelli, Fabrizio. *Le Saline di Volterra nel Granducato di Toscana.* Florence: Leo S. Olschki, 2000.

Brunelli, Giampiero. *Dizionario Biografico degli Italiani.* Vol. 52, 676–78, s.v. *Gaudenzi, Paganino.* Rome: Enciclopedia Italiana, 1999.

Caciagli, Costantino. *La casa colonica ed il paesaggio agrario nel volterrano.* Pisa: Bandecchi e Vivaldi, 1989.

Calabresi, I. *Dizionario Biografico degli Italiani.* Vol. 15, 264–68, s.v. *Buonmattei, Benedetto.* Rome: Enciclopedia Italiana, 1972.

Campanella, Tommaso. *Apologia pro Galileo.* Edited and translated by Jean-Michel Lerner. Paris: Les Belles Lettres, 2001.

Camporeale, Giovannangelo. *Gli Etruschi, storia e civiltà.* Turin: UTET, 2000.

Carafa, Paolo. "Organizzazione territoriale e sfruttamento delle risorse economiche nell'agro Volterrano tra l'orientalizzante e l'età ellenistica." *Studi Etruschi* 59 (1994): 104–21.

Casciato, Maristella, Maria Grazia Ianniello, and Maria Vitale, eds. *En-*

ciclopedismo in Roma barocca: Athanasius Kircher e il museo del Collegio Romano tra Wunderkammer e museo scientifico. Venice: Marsilio, 1986.

Casini, Bruno. "I 'Libri d'Oro' della Città di Volterra e San Miniato." *Rassegna Volterrana* LXX (1994): 391–442.

Cateni, Gabriele. *L'acropoli di Volterra: Nascita e sviluppo di una città.* Pisa: Pacini, 1981.

Cavallini, Mons. Maurizio. *Guida: Cattedrale Volterra. Battistero.* Volterra: UTA, 1957.

Cerbu, Thomas. "Melchior Inchofer, 'Un homme fin & rusé.'" In *Largo Campo di Filosofare, Eurosymposium Galileo 2001*, ed. José Mongesinos and Carlos Solís, 587–611. Las Palmas de Gran Canaria: Fundación Canaria Orotava de la Ciencia, 2001.

Cicero. *Catilinarian Orations.* Translated as *Selected Political Speeches of Cicero*, by Michael Grant. Harmondsworth: Penguin Classics, 1969.

Cipolla, Carlo M. *Cristofano and the Plague: A Study in the History of Public Health in the Age of Galileo.* London: Collins, 1973.

Cipriani, Giovanni. *Guillaume Postel, De Etruriae regionis quae prima in orbe Europaeo habitata est, Originibus, Institutis, Religione et Moribus.* Florence: Consiglio Nazionale delle Ricerche, 1986.

——. *Il mito etrusco nel Rinascimento fiorentino.* Florence: Leo S. Olschki, 1980.

Cochrane, Eric. *Florence in the Forgotten Centuries, 1527–1800.* Chicago: University of Chicago Press, 1973.

——. *Historians and Historiography in the Italian Renaissance.* Chicago: University of Chicago Press, 1981.

——. *Tradition and Enlightenment in the Tuscan Academies, 1690–1800.* Rome: Edizioni di Storia e Letteratura, 1961.

Colonna, Giovanni. "Ricerche sull'Etruria interna volsiniese." *Studi Etruschi* 41 (1973): 45–72.

Cristofani, Mauro. *Dizionario della civiltà etrusca.* Florence: Giunti Martello, 1985.

——. *La scoperta degli etruschi: Archeologia e antiquaria nel '700.* Rome: Consiglio nazionale delle ricerche, 1983.

——. "Le iscrizioni etrusche." In *Siena: Le Origini: Testimonianze e miti archeologici*, ed. Mauro Cristofani, 119–25. Florence: Leo S. Olschki, 1979.

——, ed. *Siena: Le Origini: Testimonianze e miti archeologici*. Florence: Leo S. Olschki, 1979.

Danielsson, O. E. "Annius von Viterbo über die Gründungsgeschichte Roms." In *Corolla archaeologica principi hereditario regni sueciae Gustavo Adolpho dedicata*, 1–16. Lund: Glerup, 1932.

——. *Etruskische Inschriften in Handschriftlicher Überlieferung, Skrifter utgivna av Kungliga Humanistiska Vetenskaps-Samfundet i Uppsala*, 25:3. Uppsala: Almqvist and Wiksell, 1928.

Deitz, Luc. "Die Scarith von Scornello: Fälschung und Methode in Curzio Inghiramis 'Ethruscarum antiquitatum fragmenta' (1637)." *Neulateinisches Jahrbuch* 5 (2003): 103–34.

Dempster, Thomas. *De Etruria Regali*. Florence: typis Regiae Celsitudinis apud Joannem Cajetanum Tartinum, & Sanctem Franchium, 1723–26.

De Puma, Richard D., and Jocelyn Penny Small, eds. *Murlo and the Etruscans: Art and Society in Ancient Etruria*. Madison: University of Wisconsin Press, 1994.

Devoto, Giacomo. *Tabulae Iguvinae*. Rome: typis Regiae Officinae Typographiae, 1948.

Diaz, Furio. *Il Granducato di Toscana—I Medici*. Turin: UTET, 1987.

Dini, Francesco. *Antiquitatum Etruriae, seu De situ Clanarum fragmenta historica*. Senigallia: F. A. Percineum, 1696.

——. *Dell'origine, famiglia, patron, e azzioni di Caio Mecenate gran capitano*. Venice: a spese di D. Lovisa, 1704.

Domenici, Viviano. "La ragazza rapita che conquistò il sultano." *Corriere della Sera* 3, August 2003.

Drake, Stillman. *Essays on Galileo and the History and Philosophy of Science*. Toronto: University of Toronto Press, 1999.

——. *Galileo at Work: His Scientific Biography*. Chicago: University of Chicago Press, 1978.

Edlund-Berry, Ingrid E. M. *The Gods and the Place: Location and Function of Sanctuaries in the Countryside of Etruria and Magna Graecia (700–400 B.C.)*. Stockholm: Svenska institutet i Rom, 1987.

——. "Ritual Destruction of Cities and Sanctuaries: The 'Un-found-ing' of the Archaic Monumental Building at Poggio Civitate (Murlo)." In *Murlo and the Etruscans: Art and Society in Ancient Etruria*, ed. Richard D. De Puma and Jocelyn Penny Small, 16–28. Madison: University of Wisconsin Press, 1994.

——. *The Seated and Standing Statue Akroteria from Poggio Civitate (Murlo)*. Rome: Giorgio Bretschneider, 1992.

Ehses, Stephan, ed. *Concilii Tridentini Actarum Vol. V, Pars Altera: Acta post sessionem tertiam usque ad Concilium Bononiam translatum*. Freiburg i. Breisgau: B. Herder, 1911.

Emiliozzi, Adriana. *Il Museo Civico di Viterbo: Storia delle raccolte arche-ologiche*. Rome: Consiglio Nazionale delle Ricerche, 1986.

Ernst, Heinrich, and Paganino Gaudenzio. *Ad antiquitates etruscas quas Volaterrae nuper dederunt observationes, in quibus disquisitionis astronomicae de etruscarum antiquitatum fragmentis auctor quoque no-tatur*. Amsterdam: apud Joannem Janssonium, 1639.

Everitt, Anthony. *Cicero: The Life and Times of Rome's Greatest Politician*. London: John Murray, 2001. Reprint, New York: Random House, 2003.

Fantoli, Annibale. *Galileo, for Copernicanism and for the Church* (translation by George V. Coyne of *Galileo, per il copernicanesimo e per la chiesa*). 2nd ed., rev. and corr. [Vatican City]: Vatican Observatory Publications, 1996.

Ferretti, Roberto. "La pirateria barbaresca sulle coste della Maremma." In *I Medici e lo stato senese, 1555–1609: Storia e territorio*, ed. Leonardo Rombai, 40–41. Rome: De Luca, 1989.

Finocchiaro, Maurice A., trans. and ed. *The Galileo Affair: A Documentary History*. Berkeley: University of California Press, 1989.

——. "Philosophy versus Religion and Science versus Religion: the Trials of Bruno and Galileo," in *Giordano Bruno, Philosopher of the Renaissance*, ed. Hilary Gatti, 51–96. Aldershot: Ashgate, 2002.

Fiumi, Enrico. *I confini della Diocesi Ecclesiastica del Municipio Romano e dello Stato Etrusco di Volterra*. Florence: Leo S. Olschki, 1968.

——. *L'impresa di Lorenzo de' Medici contro Volterra (1472)*. Florence: Leo S. Olschki, 1948.

——. *Volterra etrusca e romana*. Pisa: Pacini Editore, 1976.

Fowler, Murray, and Richard George Wolfe. *Materials for the Study of the Etruscan Language*. Rome: Edizioni dell'Ateneo, 1980.

Freedberg, David. *The Eye of the Lynx: Galileo, His Friends, and the Beginnings of Modern Natural History*. Chicago: University of Chicago Press, 2002.

Fumagalli, Edoardo. "Aneddoti della vita di Annio da Viterbo, O.P., 1. Annio la vittoria dei genovesi sui sforzeschi; 2. Annio e la disputa sul'Immacolata Concezione." *Archivum Fratrum Predicatorum* 50 (1980): 167–99.

——. "Un falso tardo-quattrocentesco: Lo Pseudo-Catone di Annio da Viterbo." In *Vestigia, Studi in onore di Giuseppe Billanovich*, ed. Rino Avesani, 337–83. Rome: Edizioni di Storia e Letteratura, 1984.

Galdy, Andrea. "'Con bellissimo ordine': Antiquities in the Collection of Cosimo I de' Medici and Renaissance Archaeology." Ph.D. diss., University of Manchester, 2002.

Galileo Galilei. *Dialogo sopra i due massimi sistemi del mondo tolomaico e copernicano*. Edited by Libero Sosio. Turin: Giulio Einaudi editore, 1970.

——. *Le Opere di Galileo Galilei*. Edizione Nazionale. Edited by Antonio Favaro. Florence: Barbera, 1968.

Galluccio, Ernesto. *Volterra etrusca alla luce delle nuove scoperte. Opuscula Romana: Annual of the Swedish Institute in Rome*, 2000.

Giovanelli, Fra Mario O.E.S.A. [Ordo Eremitarum Sancti Augustini]. *Cronistoria dell'antichità e nobiltà di Volterra, cominciando dal principio della sua edificazione infin'al giorno d'hoggi*. Pisa: appresso Giovanni Fontani, 1613.

Godenzi, Giuseppe. *Paganino Gaudenzi.* Bern: H. Lang/Frankfurt: P. Lang, 1975.

Gorman, Michael John. "A Matter of Faith?: Christoph Scheiner, Jesuit Censorship, and the Trial of Galileo," *Perspectives on Science* 4 (1996): 283–320.

Grafton, Anthony. *Forgers and Critics: Creativity and Duplicity in Western Scholarship*. Princeton: Princeton University Press, 1990.

Grummond, Nancy Thomson de. "Rediscovery." In *Etruscan Life and Afterlife*, ed. Larissa Bonfante, 18–46. Detroit: Wayne State University Press, 1986.

Hammond, Frederick, *Music and Spectacle in Baroque Rome: Barberini Patronage under Urban VIII*. New Haven: Yale University Press, 1994.

Harris, William V. *Rome in Etruria and Umbria*. Oxford: Oxford University Press, 1971.

Haskell, Francis. *Patrons and Painters: A Study in the Relations between Italian Art and Society in the Age of the Baroque*. Rev. 2nd ed. New Haven: Yale University Press, 1980.

Haynes, Sybille. *Etruscan Civilization: A Cultural History*. Los Angeles: The J. Paul Getty Museum, 2000.

Hebborn, Eric. *Drawn to Trouble: The Forging of an Artist: An Autobiography*. Edinburgh: Mainstream Press, 1991.

Herklotz, Ingo. *Cassiano dal Pozzo und die Archäologie des 17. Jahrhunderts*. Munich: Hirmer, 1999.

Inchofer, Melchior, S.J. *Bennonis Durkhundurkii Slavi in Spenti Accademici Sepulti Epistolam, Pro Antiquitatibus Etruscis Inghiramis: Adversus Leonis Allatii, contra easdem Animadversiones, Examen*. Cologne, Georg Genselin, 1642 [a false imprint, really Lyon].

Inghirami, Curzio. *Discorso sopra l'opposizioni fatte all'antichità toscane*. Florence: Amadore Massi and Lorenzo Landi, 1645.

——. *Ethruscarum Antiquitatum Fragmenta*. Florence: Amadore Massi, 1636 [with a false imprint of Frankfurt, 1637].

Inghirami, Lodovico. "Patriziato e cultura a Volterra in età moderna." *Rassegna Volterrana* 70 (1994): 300–305.

Johannesson, Kurt. *The Renaissance of the Goths in Sixteenth-Century Sweden: Johannes and Olaus Magnus as Politicians and Historians*, trans. James Larson. Berkeley: University of California Press, 1991.

Kaplan, Louise J. *The Family Romance of the Impostor-Poet Thomas Chatterton*. Berkeley: University of California Press, 1989.

Kircher, Athanasius. *Obeliscus Pamphilius*. Rome: Lodovico Grignani, 1650.

Lancellotti, Don Secondo. *Farfalloni de gl'Antichi Historici, Notati dall'Abate Don Secondo Lancellotti da Perugia, Accademico Insensato, Affidato, et Humorista Autore del Hoggidí.* Venice: Giacomo Sarzina, 1636.

Langford, Jerome, and Stillman Drake. *Galileo, Science and the Church.* New York: Desclee, 1966.

Leinkauf, Thomas. *Mundus combinatus: Studien zur Struktur der barocken Universalwissenschaft am Beispiel Athanasius Kirchers SJ (1602–1680).* Berlin: Akademie-Verlag, 1993.

Lerner, Lawrence S., and Edward A. Gosselin. "Galileo and the Specter of Bruno." *Scientific American* 255, no. 5 (November 1986): 126–33.

Ligota, Christopher. "Annius of Viterbo and Historical Method." *Journal of the Warburg and Courtauld Institutes* 50 (1987): 44–56.

Lisci, Niccolò Maria. *Documenti raccolti dell'illustrissimo Signor Canonico Niccolò Maria Lisci . . . intorno alle antichità Toscane di Curzio Inghirami.* Florence: Pietro Gaetano Viviani, 1739.

Lopes de Silva, A. J. *Cartas de D. Vicente Nogueira, publicadas pelo director da Biblioteca pública de Évora A. J. Lopes da Silva.* Coimbra: Impr. da universidade, 1925.

Lo Sardo, Eugenio. *Iconismi et Mirabilia da Athanasius Kircher.* Rome: Edizioni dell'Elefante, 1999.

———, ed. *Athanasius Kircher, S.J., Il Museo nel Mondo.* Rome: De Luca, 2001.

Machamer, Peter, ed. *The Cambridge Companion to Galileo.* Cambridge: Cambridge University Press, 1998.

Maetzke, Guglielmo, and Luisa Tamagno Perna, eds. *Aspetti della cultura di Volterra etrusca fra l'età del ferro e l'età ellenistica e contributi alla ricerca antropologica alla conoscenza del popolo etrusco, Atti del Convegno di Studi Etruschi ed Italici, Volterra, 15–19 ottobre 1995.* Florence: Olschki, 1997.

Maffei, Raffaele. *Commentaria Urbana.* Rome: Eucharius Silber, 1506.

Maffei, Raffaello. "Vita del Provveditore Raffaello Maffei." In *Storia Volterrana del Provveditore Raffaello Maffei,* ed. Annibale Cinci, VII–LX. Volterra: Tipografia Sborgi, 1887.

Magnus, Olaus. *Historia de gentibus septentrionalibus, earumque diversis statibus, conditionibus, moribus, ritibus, superstitionibus, disciplinis . . . Authore Olao Magno Gotho*. Rome: apud Ioannem Mariam de Viottis parmensem, 1555.

Magnuson, Torgil. *Rome in the Age of Bernini*. Stockholm: Almqvist and Wiksell, 1982 [vol. I], 1986 [vol. II].

Marrucci, Angelo. "I bagni e le moie del volterrano alla metà del XVII secolo." *La comunità di Pomarance* 11 (1997): 28–31, 37–41.

——. "Nützliche Metalle: Steinsalz und Silber." In *Otto der Große und Europa: Volterra von Otto I bis zur Stadtrepublik*, ed. Andrea Augenti, 66–72. Siena: Nuova Immagine, 2001.

——. *Vol. III: I personaggi e gli scritti: Dizionario biografico e bibliografico di Volterra*, in *Dizionario di Volterra*, ed. Lelio Lagorio. Pisa: Ospedaletto, 1997, 1056–59, s.v. *Inghirami, Curzio*.

Martelli Cristofani, Marina. "MS Sloane 3524." In *Siena: Le Origini: Testimonianze e miti archeologici*, ed. Mauro Cristofani, 136–43. Florence: Leo S. Olschki, 1979.

Massa-Pairault, Françoise-Hélène. "La stele di 'Avile Tite' da Raffaele il Volterrano ai giorni nostri." *Mélanges de l'École Française de Rome: Antiquité* 103 (1991): 499–528.

McPhee, Sarah. *Bernini and the Bell Towers: Architecture and Politics at the Vatican*. New Haven: Yale University Press, 2002.

Miller, Peter N. *Peiresc's Europe: Learning and Virtue in the Seventeenth Century*. New Haven: Yale University Press, 2000.

Montagu, Jennifer. *Roman Baroque Sculpture: The Industry of Art*. New Haven: Yale University Press, 1998.

Morolli, Gabriele. *"Vetus Etruria": Il mito degli Etruschi nella letteratura architettonica nell'arte e nella cultura da Vitruvio a Winckelmann*. Florence: Alinea Editrice, 1985.

Musto, Ronald G. *Apocalypse in Rome: Cola di Rienzo and the Politics of the New Age*. Berkeley: University of California Press, 2003.

O'Malley, John W. *The First Jesuits*. Cambridge: Harvard University Press, 1993.

Pallottino, Massimo. "Héritages lexicaux." In *Les Étrusques et l'Europe*, 246–47. Milan: Fabbri, 1992.

———. *Testimonia Linguae Etruscae.* 2nd ed. Florence: La Nuova Italia, 1968.

———. *Thesaurus Linguae Etruscae.* Vol. I. Rome: Consiglio Nazionale delle Ricerche, Centro di Studio per l'Archeologia Etrusco-Italica, 1968.

Parise, Nicola. *Dizionario Biografico degli Italiani.* Vol. 15. Rome: Istituto dell Enciclopedia Italiana, 1972, 145–47, s.v. *Buonarroti, Filippo.*

Parkes, M. B. *Pause and Effect: An Introduction to the History of Punctuation in the West.* Aldershot: Scolar Press, 1992.

Pastine, Dino. *La nascita dell'idolatria: l'Oriente religioso di Athanasius Kircher.* Florence: La Nuova Italia, 1978.

Paturzo, Franco. *Mecenate, il ministro di Augusto: Politica, filosofia, letteratura nel periodo augusteo.* Cortona: Calosci, 1999.

Phillips, Kyle M., Jr. *In the Hills of Tuscany: Recent Excavations at the Site of Poggio Civitate.* Philadelphia: University Museum, University of Pennsylvania, 1993.

Phillips, Kyle M., Jr., and Erik O. Nielsen. "Poggio Civitate." In *Case e Palazzi d'Etruria*, ed. Simonetta Stopponi, 64–69. Milan: Electa, 1985.

Portoghesi, Paolo. *Francesco Borromini: Architettura come linguaggio.* Milan: Electa, 1967.

Postel, Guillaume. *De Etruriae regionis quae prima in orbe Europaeo habitata est, Originibus, Institutis, Religione et Moribus . . . et imprimis De Aurei Saeculi Doctrina commentaria.* Florence, 1551.

Pozzi, Giovanni, and Lucia Ciapponi. *Hypnerotomachia Poliphili.* Padua: Antenore, 1980.

Redondi, Pietro. *Galileo Eretico.* Turin: Einaudi, 1983.

Reeves, Marjorie. *The Influence of Prophecy in the Later Middle Ages: A Study in Joachimism.* Oxford: Clarendon Press, 1969.

———. *Joachim of Fiore and the Prophetic Future.* London: SPCK, 1976.

———, ed. *Prophetic Rome in the High Renaissance Period: Essays.* Oxford: Clarendon Press, 1992.

Renieri, Vincenzo. *Monopanthon, De Ethruscarum Antiquitatum fragmentis Scornelli prope Vulterram repertis Disquisitio Astronomica.* Florence: Amadore Massi, 1638.

Repgen, Konrad. *Dreißigjähriger Krieg und Westfälischer Friede: Studien und Quellen*. Paderborn: Ferdinand Schöningh, 1998.

Rochon, André. *Formes et significations de la "Beffa" dans la littérature italienne de la Renaissance: Études réunies*. Paris: Université de la Sorbonne Nouvelle, 1972–75.

Rodén, Marie Louise. *Church Politics in Seventeenth-Century Rome: Cardinal Decio Azzolino, Queen Christina of Sweden and the Squadrone Volante*. Stockholm: Almqvist & Wiksell, 2000.

Roncalli, Francesco, ed. *Scrivere Etrusco*. Milan: Electa, 1985.

Rowland, Ingrid D. *The Culture of the High Renaissance: Ancients and Moderns in Sixteenth-Century Rome*. Cambridge: Cambridge University Press, 1998.

——. "Due 'traduzioni' rinascimentali dell'*Historia Porsennae*." In *Protrepticon: Studi in memoria di Giovannangiola Secchi Tarugi*, ed. Sesto Prete, 125–33. Milan: Istituto di Studi Umanistici Francesco Petrarca, 1989.

——. "Early Attestations of the Name 'Poggio Civitate.'" In *Murlo and the Etruscans: Art and Society in Ancient Etruria*, ed. Richard D. De Puma and Jocelyn Penny Small, 3–5. Madison: University of Wisconsin Press, 1994.

——. *The Ecstatic Journey: Athanasius Kircher in Baroque Rome*. With an introduction by F. Sherwood Rowland. Chicago: Department of Special Collections, University of Chicago Library, 2000.

——. "Etruscan Inscriptions from a 1637 Autograph of Fabio Chigi." *American Journal of Archaeology* 93 (1989): 423–28.

——. "Etruscan Secrets." *New York Review of Books*, July 5, 2001.

——. "Il mito di Porsenna: Leggenda e realtà." In *Il Mito nel Rinascimento*, ed. Luisa Rotondi Secchi Tarugi, 391–407. Milan: Franco Cesati Editore, 1993.

——. "*L'Historia Porsennae* e la conoscenza degli Etruschi nel rinascimento." *Res Publica Litterarum* 9 (1989): 185–93.

——. "Th' United Sense of the Universe: Athanasius Kircher at Piazza Navona," *Memoirs of the American Academy in Rome* 46 (2001): 153–81.

Sallust (Gaius Sallustius Crispus). *The Conspiracy of Catiline*. Translated by S. A. Handford. Harmondsworth: Penguin Classics, 1963.

Santillana, Giorgio de. *The Crime of Galileo*. Chicago: University of Chicago Press, 1955.

Schiebe, Marianne Wifstrand. "Annius von Viterbo und die schwedische Historiographie des 16. und 17. Jahrhunderts." *Skrifter Utgivna av Kungliga Humanistiska Vetenskaps-Samfundet i Uppsala (Acta Societatis Letterarum Humaniorum Regiae Upsaliensis)* 48 (1992): 7–26.

Scipione Ammirato the Younger [Cristoforo Bianco], ed. *Istorie fiorentine di Scipione Ammirato con l'aggiunte di Scipione Ammirato il Giovane* (Florence: Amadore Massi, 1647).

Scott, John Beldon. *Images of Nepotism: The Painted Ceilings of Palazzo Barberini*. Princeton: Princeton University Press, 1991.

Small, Jocelyn Penny. *Cacus and Marsyas in Etrusco-Roman Legend*. Princeton: Princeton University Press, 1982.

Solaini, E. "Il falso 'Estratto del Camerotto di Volterra.'" *Rassegna Volterrana* I (1924): 17–19.

Solinas, Francesco, ed. *Cassiano dal Pozzo: Atti del Seminario Internazionale di Studi*. Rome: De Luca, 1989.

Lo Spento Accademico Sepolto. *Lettera Sopra il libro intitolato Leonis Allatii Animadversiones in Ethruscarum Antiquitatum Fragmenta*. Florence: Amadore Massi and Massi and Lorenzo Landi, 1641.

Steingräber, Stefan. *Città e necropoli dell'Etruria: Luoghi segreti e itinerari affascinanti alla riscoperta di un'antica civiltà italica*. Rome: Newton Compton Editori, 1983; translation by Giorgio Ziffer of *Etrurien: Städte, Heiligtümer, Nekropolen*. Munich: Hirmer Verlag, 1981.

Steingräber, Stefan, and Horst Blanck, eds. *Volterra: Etruskische und mittelalterliches Juwel im Herzen der Toskana*. Mainz: Philipp von Zabern, 2003.

Stephens, Walter E. "Berosus Chaldaeus: Counterfeit and Fictive Editors of the Early Sixteenth Century." Ph.D. diss., Cornell University, 1979.

——. "The Etruscans and the Ancient Theology in Annius of

Viterbo." In *Umanesimo a Roma nel Quattrocento*, ed. Paolo Brezzi
and Maristella Panizza Lorch, 309–22. New York: Barnard College, 1984.

——. *Giants in Those Days: Folklore, Ancient History, and Nationalism.*
Lincoln: University of Nebraska Press, 1989.

Stoltzenberg, Daniel, ed. *The Great Art of Knowing: The Baroque Encyclopedia of Athanasius Kircher.* Stanford: Stanford University Li-
braries/Fiesole: Cadmo, 2001.

Taylor, Donald. *Thomas Chatterton's Art.* Princeton: Princeton University Press, 1978.

Terrenato, Nicola, and Alessandra Saggin. "Ricognizioni archeologiche nel territorio di Volterra." *Archeologia Classica* 46 (1994):
465–82.

Tigerstedt, E. N. "Ioannes Annius and *Graecia mendax*." In *Classical,
Medieval, and Renaissance Studies in Honor of Berthold Louis
Ullmann*, ed. Charles Henderson, vol. 2, 293–310. Rome: Edizioni
di Storia e Letteratura, 1964.

Tyson, Gerald, and Sylvia Wagonheim, eds. *Print and Culture in the
Renaissance: Essays on the Advent of Printing in Europe.* Newark:
University of Delaware Press, 1986.

Vasari, Giorgio, *Le Vite dei più eccellenti pittori, scultori ed architettori.*
Edited by Rosanna Bettarini. Florence: Studio per Edizioni
Scelte, 1986.

Vulcanius, Bonaventura. *De literis et lingua Getarum.* Louvain: Plantin, 1597.

Waddy, Patricia. *Seventeenth-Century Roman Palaces: Use and the Art of
the Plan.* New York: Architectural History Foundation/Cam-
bridge: MIT Press, 1990.

Walker, D. P. *The Ancient Theology: Studies in Christian Platonism from
the Fifteenth to the Eighteenth Century.* London: Duckworth, 1972.

Weiss, Roberto. "Traccia per una biografia di Annio da Viterbo."
Italia medioevale e umanistica 5 (1962): 425–41.

——. "An Unknown Epigraphic Tract by Annius of Viterbo." In *Italian Studies Presented to E. R. Vincent*, ed. E. R. Vincent, K. Forster,
and U. Limentani, 101–20. Cambridge: Heffer, 1962.

Wilding, Nick. "Writing the Book of Nature: Natural Philosophy and Communication in Early Modern Europe." Ph.D. diss., European University Institute, 2000.

Zamarchi Grassi, Paola, and Dario Bartoli. *Museo archeologico nazionale G. Cilnio Mecenate, Arezzo*. Rome: Istituto poligrafico e Zecca dello Stato, Libreria dello Stato, 1993.

Index

Page numbers for illustrations are in italic.